THE MISCHIEF OF MATH

SHORT STORIES OF CLOWNS, CONTORTIONISTS, AND COURT-JESTERS

THE MISCHIEF OF MATH

SHORT STORIES OF CLOWNS, CONTORTIONISTS, AND COURT-JESTERS

INAVAMSI ENAGANTI
Param Innovation Centre, India

NIVEDITA GANESH
New York University, USA

BUD MISHRA
New York University, USA

Illustrated by
ALEXANDER LU

NEW JERSEY • LONDON • SINGAPORE • BEIJING • SHANGHAI • HONG KONG • TAIPEI • CHENNAI • TOKYO

Published by

World Scientific Publishing Co. Pte. Ltd.

5 Toh Tuck Link, Singapore 596224

USA office: 27 Warren Street, Suite 401-402, Hackensack, NJ 07601

UK office: 57 Shelton Street, Covent Garden, London WC2H 9HE

Library of Congress Cataloging-in-Publication Data
Names: Enaganti, Inavamsi, author. | Ganesh, Nivedita, author. | Mishra, Bud, author. |
    Lu, Alexander (Illustrator), illustrator.
Title: The mischief of math : short stories of clowns, contortionists, and court-jesters /
    Inavamsi Enaganti, Nivedita Ganesh, Bud Mishra ; illustrated by Alexander Lu.
Description: New Jersey : World Scientific, 2024.
Identifiers: LCCN 2024002422 | ISBN 9789811287596 (hardcover) |
    ISBN 9789811288029 (paperback) | ISBN 9789811287602 (ebook) |
    ISBN 9789811287619 (ebook other)
Subjects: LCSH: Mathematics--Fiction. | Logic--Fiction. | LCGFT: Short stories.
Classification: LCC PN6071.M3 E53 2024 | DDC 808.8/0356--dc23/eng/20240226
LC record available at https://lccn.loc.gov/2024002422

British Library Cataloguing-in-Publication Data
A catalogue record for this book is available from the British Library.

For any available supplementary material, please visit
https://www.worldscientific.com/worldscibooks/10.1142/13711#t=suppl

Well, the way of paradoxes is the way of truth.
To test reality we must see it on the tightrope. When
the verities become acrobats, we can judge them.

Oscar Wilde

PREFACE

Years later, one of us would remember those wee hours of a beautiful spring morning — having just woken up from a strange dream, in which one finds oneself lying naked with strange letters and numbers tattooed on one's back. One rushes out an email to cognoscenti (geneticists, bioinformaticists, epidemiologists, immunologists, virologists, mathematicians, etc.) wondering if the tattoo in the dream spells out some **HLA** connection with Covid that classified patients into symptomatic or asymptomatic. Little did anyone suspect that the truth would remain slippery, vacillating among conjectures, dreams, nightmares, conspiracy theories and hypotheses, some not even wrong.

A group, seeking to falsify many of these hypotheses with minimal viable (in Silico) experiments, emerges, with the group members gingerly walking on the conjectures' tight ropes like verities' acrobats. The group realized that, as Wilde said, "the way of paradoxes is the way of truth. To test reality [they] must see it on the tightrope. When the verities become acrobats, [they] can judge them." Originally christened Covfefe (Covid For Ever, For Everyone), the group changed its name to RxCovea and continued to grow as it sought various solutions.

RxCovea was original in three ways: it was suspicious of the truth, prima facie, based on data and information as provided by the policy makers (Court-Jesters); it tried to vet out the truth promoted in social media by the conspiracy theorists (Clowns); and it also tried to sift the truth that was not even wrong and consisted of un-understandable pseudo-science supported by unicorn technologists (Contortionists). The group trained young scientists to prepare themselves as problem-solvers dealing with an ever-evolving whack-a-mole of wicked problems (problems with many interdependent factors making them seem impossible to solve).

Two of the young mentees and an aging mentor thought it would be great to recreate RxCovea's pedagogy in a widely available framework, first in the form of a paper (INVENTIONS OF INTERVENTIONS: Data Driven Strategies In Pandemic Research And Control) published in the National Academy of Inventors Journal, and then in form of a book — which is what you are holding as you prepare for your own wicked problems amidst the cacophony and crowds of clowns, contortionists and court jesters.

Hope you do better!
Inav, Nivedita and Bud

CONTENTS

ACKNOWLEDGEMENTS

We extend our gratitude to those who have come before us, for providing sturdy shoulders to stand upon. To those among the living, thank you for your support without triggering any adverse Butterfly Effects. And to those yet to be born, we appreciate your decision not to invent a time machine and halt our book-writing escapades.

And thanks most of all to our families, friends and fools.

Part 1

The Archipelago of Serene Deep

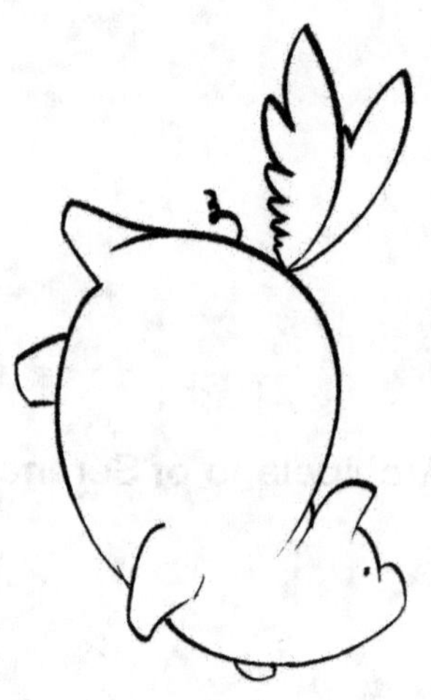

WHEN PIGS FLY

The first batch of pigs arrives from overseas onto a remote island.

On an earth far, far away in an island nation that had been isolated for hundreds of years from the rest of their world, the people were cheering for change because their troubles were soon to be over. The newly installed regime had just opened up the island nation's borders for trade and travel. This nation had isolated itself from external culture, food, goods, technology, and ideas for eons, so this meant preparing for an influx of new ideas and changes.

The rest of the world was speeding ahead with great scientific endeavors, and this little nation wanted so dearly to catch up. Young and presumptuous Don, a prominent scientist of the nation had, like the rest of the people of the land, never laid eyes on the pink, furry creature which others called a pig. In this very normal but scientifically defunct land, we begin our story with a newly imported batch of pigs on a quaint little farm.

As per the nation's regulations, all new creatures had to go through a nutrient injection that helped cleanse their bodies of foreign bacteria and toxins. The farmers were not equipped to understand these procedures and reached out to the nearby research lab for help.

At the lab, the scientists were in a fix. They had just received the call for help from the farm, asking them to provide the most appropriate formulation to cleanse the newly imported creatures. The standard nutrient injections made bones heavier and stronger. However, if these are injected into a flying creature, the creature would be prevented from flying. So, they had a special nutrient injection that was much more expensive as it was specially designed for flying creatures. Now, the question was if this new pig creature could fly.

The extant native island creatures had always been there, and the humans of this land already knew which ones could fly and which couldn't. But the introduction of this mysterious animal

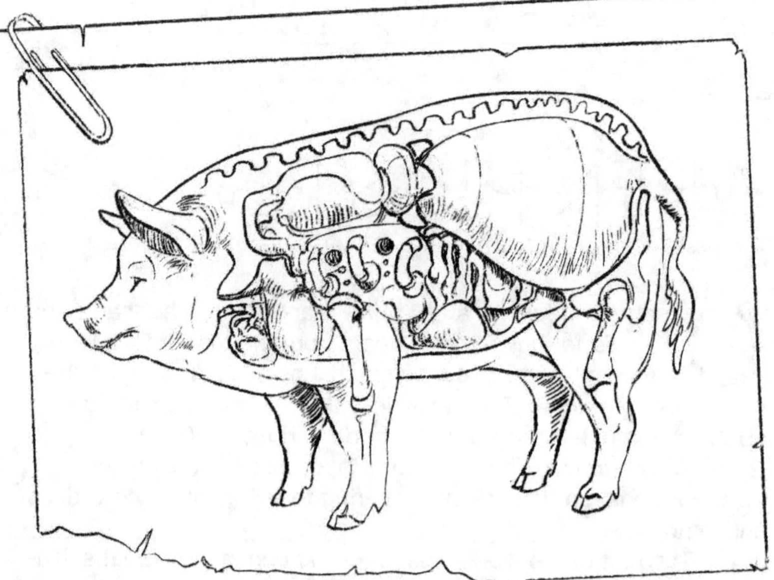

raised a new question – "How do you determine if a creature can fly?" Obviously, they could keep the creatures under observation for a long enough time to verify if they ever flew, but they did not have time on hand.

So the scientists pulled up some data on common creatures. They looked at cows, dogs, pigeons, cats, crows, goats, eagles and many more. Upon analyzing their biological specifications, they made a breakthrough. They devised and quickly patented what they called a complex **machine learning algorithm**: if the bone density was lower than a certain threshold, then the animals could fly. If it was higher, they could not fly.

*This is a common misinterpretation of **causal relation**. Flying creatures have lower bone density and not the other way around.*

The head researcher, Don, personally oversaw this important initiative.

"I really don't want to make a mistake. So to be safe, let's ask for the pigs with the densest bones. If the densest pig bones are below the threshold, then it follows that they can fly," Don haughtily declared, convinced by his faulty logic.

Don headed to the farm to choose a sample of pigs. He

summoned the local farmers, stout, tanned Bob and fair, lanky Chip, and demanded, "Get me the pigs with the densest bones. I need to take them back to the lab and do a few tests before giving you the correct nutrient injection."

A mystified Bob replied, "I don't have the faintest clue as to which pigs have the densest bones!"

Don replied, "Get the pigs here; we can figure it out together."

Just as Bob was about to round up the pigs, Chip remembered that the importers had warned him it was currently pig mating season. So he instantly jumped in saying, "Mister, it is mating season and sending pigs of both genders will create a big mess, so decide on either male or female pigs."

Don was getting frustrated as it was a hot, sunny day and he was sweating profusely. He had reckoned that this trip would be short, but with these two not-so-bright farmers, he wasn't so sure anymore. Don needed dense bones, and these guys seemed clueless! As the sweat streamed down his face, he exclaimed "Dang this sun." and then with a flash, "Wait, the sun! Calcium! Tan! I am a genius!"

Don turned to the farmers and proclaimed, "The darker pigs probably get more sunlight and thus have more calcium in their bones. Go find out which gender is darker, as it will be the gender with the denser bones. I need a survey. Bob, go to the eastern side and tell me how many male and female pigs are darker than the back of your palm. Chip, you head to the western side and get me a count quickly."

Some time later, with their analyses in hand, Chip returned with a huffing and puffing Bob lagging behind. Bob found that 27% of the male pigs were darker than his hand compared to only 25% of female pigs. Chip found that 53% of male pigs and 50% of female pigs were darker than his hand. So, they both reasoned that the candidates for the lab should be chosen from the male pigs.

There was a reason, however, that Don was the scientist, and it was precisely now that he chose to show his true colors. With a superior snort he began, "That's all very well, but with my advanced science degree, one thing I did learn is that aggregating test results yields better approximations."

Chip and Bob looked at each other as if they shared a brain. With the fancy words he was using, they figured he must be right, and both nodded, grateful to have such a learned guide.

"Alright, taking the aggregate we see that 40% of the females passed the tests, and 37% of males passed the tests, making our clear winner the females!"

"That sounds weird. How can males fare better in both tests individually and yet females fare better overall?" Chip inquired, sure that he was missing something.

Don sniffed once more, "Chippy boy, you must learn, the numbers don't lie! We are men of science, not barbarians. Let us fight this with logic. Go get the female pigs."

But something certainly was amiss in Don's "foolproof" logic. A closer inspection of the numbers in the following table tells us why:

	Female	Male
Test 1: Bob Darker than hand pigment	25% (10/40)	**27%** (16/60)
Test 2: Chip Darker than hand pigment	50% (30/60)	**53%** (21/40)
	40% **(40/100)**	37% (37/100)

Despite both tests concluding that more males had darker skin, it appeared that more females have darker skin when you consider the aggregate. This transpired because there was a disproportionate allocation of pigs by their gender into each test i.e. the number of males and females in each test were different (40 or 60). The reason Bob tended to have a lower percentage of pigs that were darker than his hand pigment was that he was much more tanned than Chip, so the test on his set of pigs was more stringent.

*Classic **Simpson's paradox!***

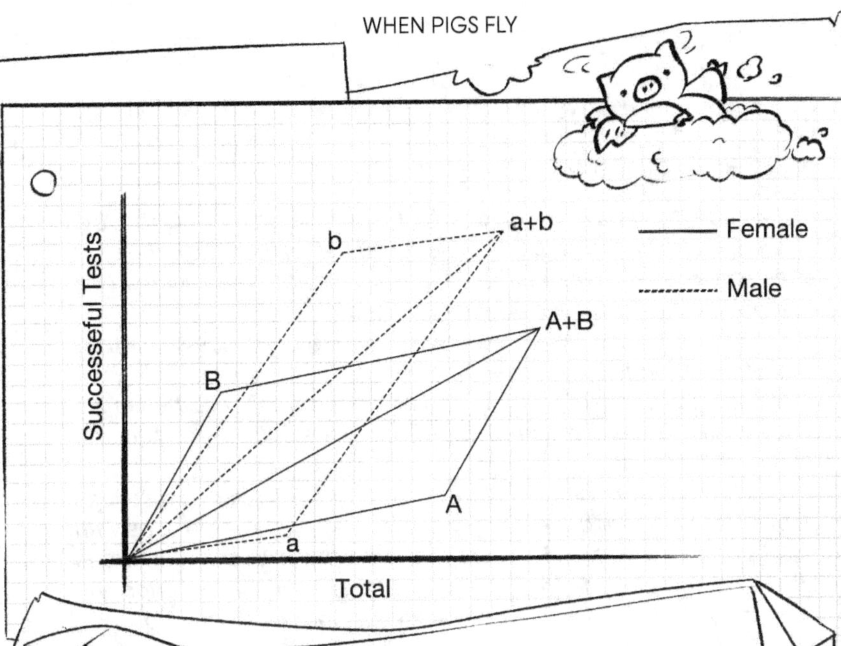

*Intuition behind **Simpson's paradox**: this is a plot between the total number of trials and the number of them that are successes. The greater the angle with the x-axis, the greater the success rate. Despite males (red) being better on both group A (Test 1) and group B (Test 2) individually, when combined we see that females (blue) seem to be better.*

Now that supposedly the desired gender had been chosen, Don announced that they would find the densest boned female pigs and send them to the research lab. Well, he could go around and try comparing each pig to find the heaviest ones, but that would take too long. Thinking back to stout Bob lying in the corner, exhausted after coming in behind Chip, Don concluded that the densest boned pigs would be heavier and thus slower.

Heavier pigs could just have more fat or larger/heavier bones and do not in fact indicate anything about bone density.

"Chip, Bob, scare the pigs and catch the slowest ones!" Don commanded.

Although a bit confused, both Chip and Bob decided to put their trust in science and listen to Don. Little did Don know that his half-baked ideas would just make everything worse!

Chip clanged two metal plates together and startled the pigs who scurried every which way. They caught the stragglers and bundled them into the truck to be sent off.

Don thought that pigs with heavier bones would be the slowest, but weak-boned pigs also run slowly. When they chose to chase the pigs and pick up the stragglers, they forgot to account for osteoporosis (a low-density bone condition).

In reality, the female pigs were fairer than males since they were housed indoors to breed piglets and received too little sunlight. As a consequence, many had osteoporosis. Thus, the female pigs with osteoporosis ran too slow and were caught and brought to the research lab.

Not only did the farm choose the lighter-skinned female pigs with relatively lower bone density but they also chose the ones with osteoporosis, which led to the pigs with the lowest of low bone density being picked. This was diametrically opposite to choosing the pigs with the highest bone density, as originally intended!

Some science.

Back at the research lab, the eager scientists dissected a few pigs and carefully retrieved their bones. They prepared to go ahead and find the bone density.

This island nation's bone density scanner could only contrast light colors well. Furthermore, this newly developed machine could accommodate only smaller bones with lower weight. Thus, most of the pigs' bones were not suitable. After a bit of sorting, a set of bones were chosen.

They inevitably chose bones with even lower calcium and focused on the smaller ones that would fit comfortably in the scanner and not weigh too much. This approach leads to a classic example of **survivorship bias***, which is the logical error of concentrating on a particular subset of your sample population (bones) that made it past some selection process (what could be optimally scanned) and overlooking those that did not. Thus, the results are representative of the survivor bones that made their way into the machine and not the entire set of bones.*

Ding ding! The bone density scanner chimed as it completed its analysis. Don strolled over and picked up the result sheet.

With the results, they used their state-of-the-art algorithm that they had recently discovered. Don exclaimed, "Well, it's lower than the threshold, so I guess pigs can fly!"

Reflecting back on the algorithm: If bone density was lower than a certain threshold, then the animal could fly. If it was higher, then it could not fly.

An animal's ability to fly might indicate a lower bone density, but does that mean that lower bone density indicates that an animal is more likely to fly? Hmm, sounds an awful lot like **prosecutor's fallacy** *at play, when P(A|B) is confused with P(B|A) which is related by* **Bayes' theorem** *as shown below.*

$$P(A|B) = \frac{P(B|A) * P(A)}{P(B)}$$

A, B = events

P (A | B) = probability of event A occurring given that event B is true

P (B | A) = probability of event B occurring given that event A is true

P (A), P (B) = marginal probabilities of event A and event B

Here:
A = Animal can fly
B = Animal has bone density lower than a threshold

Armed with his supposed truth and the nutrient injection for flying creatures, Don headed back to the farm to inject all the pigs with nutrients and complete his mission.

Zoink! As he pulled out the needle, Don turned to Chip and Bob and said, "That's the last one. Take care that they do not fly away, so keep them indoors."

The perennially confused Chip replied, "I have never seen them fly. Are you sure they can?"

Don wore his superior smirk with his head held high and stated, "Just because you don't observe something doesn't mean it doesn't happen when you are not observing. Don't fall prey to such obvious loopholes."

Then, Don made his way back to the lab, eagerly waiting to announce to the island's journalists that pink furry pigs can fly.

As the night neared, Chip and Bob drove the pigs indoors as they feared that they would fly away. Before closing the door, Chip picked up a huge tank of water along with a huge box of feed and set it up on a tall shelf next to the wall. Chip turned to Bob and said, "Since they can fly anyway, it's better to keep the food up here so that it is easier to clean."

As the days rolled by, the feed and water dwindled as rats and mice fed on this feast rather than the pigs. The pigs, on the other hand, grew weak, frail and slow.

Chip and Bob were confused as nothing seemed out of the ordinary. Every night they diligently refilled the feed and water, which would be empty by the morning, so they ruled out the food as the issue. Since the pigs had recently received the nutrient injection, there was a chance that the pigs' condition was a side effect, and thus they decided to call Don.

Don decided to split the pigs into two groups to conduct some tests. He asked Bob and Chip to weigh all the pigs and record their weight. Once done, he put the 50 heaviest pigs in Room 1 and the 50 lightest pigs in Room 2.

The curious Chip asked, "Honorable sir, why must we weigh the pigs?"

Don replied, "I have a new health pill I want to test on them, but to judge if they are recovering, I need to see if their weights increase."

Chip understood and after a while came back. "Sir Don, didn't you say last time that averaging or aggregating is better than individual test results?"

Don turned around to hide his embarrassment and thought, "Why didn't I think of that? This illiterate guy is just mimicking what I said before," and then said out loud,

"Hehe! I was just testing you. Nice catch. Once you're done, just record the average weight of the groups."

"Yes, sir!" both Chip and Bob replied.

The next day, Don visited Room 1 with the heavier pigs and found it to be too crowded, so he made Bob transfer a few pigs to Room 2.

As the stout Bob was a little lazy, he picked up the lighter pigs in Room 1 and moved them to Room 2.

The following day, Don visited to record results. On weighing the pigs in both rooms, he found that the average weights in both rooms had gone up. Satisfied, the three men heaved a sigh of relief.

Clearly, the average weight of both rooms will increase even though every single pig in both rooms lost weight. They forgot that Bob moved some pigs from Room 1 to Room 2 and he chose the lightest of pigs in Room 1.

Alas, they fell prey to **Will Rogers phenomenon**.

	Room 1		Room 2	
	Weights	Averages	Weights	Averages
Initial Measurement	10, 12 14, 16	13	5, 6, 7	6
After Bob's Movement	14, 16	15	5, 6, 7 10, 12	8
Measurement After Pigs Lose Weight	13, 15	14	4, 5, 6 9, 11	7

Chip was a little worried that the pigs looked even weaker even though the weight of both rooms had increased, but he decided to keep quiet, knowing that Don would put him down with some intelligent rebuke.

To confirm his fears, when he opened the doors the next day, he found that a couple of pigs were dead. He was at a loss for words, and that's when it occurred to him that the reason they were weak was maybe because they could not fly freely in the sky. He opened the doors and pointed to the sky, gesturing wildly, but none of the pigs seemed to care.

He then remembered what mother birds do to the noncooperative younglings to make them fly – they push them off heights to give them a flying start. So, he picked up his phone and scrolled till he found the farm's cargo plane pilot and placed a call.

The next day, the nearby town square was in upheaval. The front page of the local newspaper carried a picture of a pink furry creature called a pig and a quote from a handsome and scholarly researcher named Don cheerfully stating "Pigs can fly."

The polarized townsmen were quarreling when a little boy looked up and pointed as he heard the sound of a plane. He shouted, "Mama, pink flying!"

Like deer startled by a noise, suddenly everyone looked up.

An old man grumbled, "Well what do you know, those pigs really can fly," and continued in alarm, "but why are they coming in so fast?"

1.2
EXERCISES

1. Create two groups of people with scores for each person. Ensure the following conditions. Mathematically show that both conditions are satisfied. (Hint: **ecological fallacy**)

   a. On average, Group A has higher score than group B
   b. If we pick a random person from group A and a random person from group B then 90% of the time, the person from group B will have a higher score than group A.

2. Drug or no drug?

   You are investigating the effectiveness of a drug against a deadly disease. You are given access to data collected by health insurance companies about their customers. You divide the diseased customers into two groups: those that took the drug, and those that didn't take the drug. Some of the customers recovered, others unfortunately didn't recover.[1]

	Recovery	No Recovery	Total	Recovery Rate
Drug	20	20	40	
No Drug	16	24	40	
Total	36	44	80	

   a. Calculate the recovery rates (in %) for both groups ("drug" or "no drug").

   b. If you were diseased, would you take the drug, or not?

[1] Exercises MLSS 2019: Causality, Joris Mooij, August 27, 2019

Upon closer inspection of the data, you notice something peculiar when you group patients according to gender.

c.   Calculate the recovery rates (in %) for both groups ("drug" or "no drug") for each subpopulation (males and females) separately.

Male	Recovery	No Recovery	Total	Recovery Rate
Drug	18	12	30	%
No Drug	7	3	10	%
Total	25	15	40	

Female	Recovery	No Recovery	Total	Recovery Rate
Drug	2	8	10	%
No Drug	9	21	30	%
Total	11	29	40	

d.   In light of the numbers, would you take the drug if you were diseased or not?

e.   What would be your advice to a diseased patient with unknown gender?

3. Monty Pigs

   a. There are three rooms with the following pigs in each. Room
      One has two black pigs, Room Two has two white pigs, and
      Room Three has one white and one black pig. If you randomly
      chose a room and pulled out a pig and found it to be black,
      then what is the probability that the other pig in the room is
      white?

   b. There are four rooms with the following pigs in each. Room One
      has three black pigs, Room Two has two black and one white
      pigs, Room Three has one black and two white pigs, and Room
      Four has all white pigs. If you randomly chose a room and pulled
      out a pig and found it to be black, then what is the probability
      that there is at least one other black pig in the room?

4. Which of the following statements are true or false? Analyze based on
   causality and give a reason for each answer.

   a. More ice cream makes the weather warm.

   b. Warm weather results in more ice cream sales.

   c. Students that spend more time watching TV tend to receive
      lower exam scores. This means that watching TV causes poor
      academic performance.

   d. Most heroin addicts used marijuana before they tried heroin.
      Clearly, marijuana is a gateway drug that causes heroin abuse.

LO☐K ME ☐P

Topic	Terms
Scientific Method	Hypothesis, Control Group, Dependent/Independent Variable, Peer Review, Falsifiable, Reproducible
Error	Error Propagation, Propagation of Uncertainty, Type 1 Error, Type 2 Error, False Discovery Rate (FDR), Overfitting, Underfitting
Measure	Precision, Recall, F1 Score, Accuracy, Specificity
Bayes Theorem	Likelihood, Prior, Posterior
Causality	Correlation, Causal Inference, Causal Graphs, Counterfactual Thinking, Probabillity Raising and Temporal Priority, Prima Facie Causality, Genuine and Spurious Causality
Other Relevant Paradoxes	Lindley's Paradox, Friedman's Paradox, Stein's Paradox

NOTES

ULTRA-ALTRUISM

One small kind act starts a chain reaction.

In this dog-eat-dog world, no one would blink twice at an injustice, even if it meant that they could have prevented it. Cruel, merciless, and capitalistic, these people would rather sacrifice a limb than fall behind in the rat race.

This stance hints at a proud meritocracy that values only results and nothing else, nothing to remind them of their humanity. This cut-throat culture eroded all semblance of kindness and selfless actions. Everyone was in it for themselves.

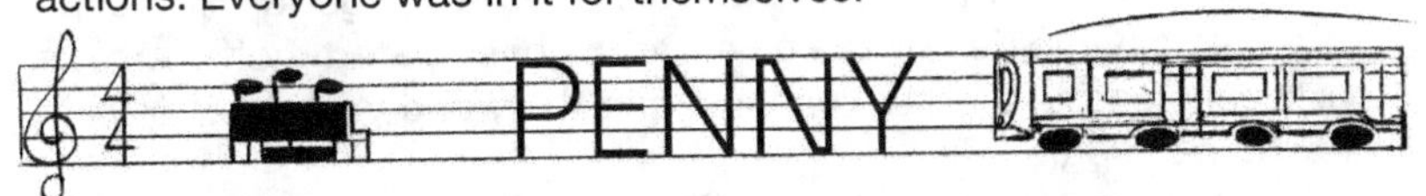

"I can't believe I don't have enough change for this stupid train ticket!" Penny exclaimed.

Rummaging deep into her pockets and the corners of her bag, she searched for just 20 pence more to buy her ticket. Frustrated, she emptied the contents of her bag onto the floor, flicking through the detritus and having no luck.

This timing was so terrible! She was on the way to her orchestra's big show, her first performance as lead cellist and time was running out.

As passengers flashed by, not giving a second glance to the hullabaloo on the floor, Penny grew increasingly agitated. She didn't bother to ask for help. She knew she would never bother to stop if the situation was reversed. How would that help her?

She was dejectedly walking back when a glint of light hit her eyes. There was a coin laying in a crack in the sidewalk. 20 pence – just what she needed! A random act of kindness, an act of God as it were.

She bought the ticket and made it in time for the show, what luck! Little did she know that a talent recruiter from the philharmonic was in the audience and took note of her rousing Dvorak cello solo.

The next day, the talent recruiter called Penny and told her that they were interested in her. Putting the phone down, she pondered – "What if I had never found the coin?" Her thoughts wandered as she imagined a world where one of the countless passersby threw her a coin, a selfless kind deed — something unheard of in her world. What if she had to get to the hospital to donate blood for her mother's critical condition? Or, what if it was her last day to pay rent and her landlord was unwilling to extend $50 of credit that she promised to pay the next day and instead evicted her? She wouldn't be so lucky every time. She needed to change things for the better. As her competitive spirit waned, compassion rose in its place.

"What could make someone in this selfish country look beyond their own needs and help someone else for once? Why is there never a good Samaritan?"

She thought and she thought. Even in poetry or art, their entire culture was plagued with stories of celebrating merit and punishing mediocrity, all in the name of ambition and pursuing power.

She couldn't think of a single tale of hope and good will. Maybe, she would have to set an example. Be the one to start it off. Perhaps, that would lead to a chain reaction, and the world would not walk by as people struggled.

So began the grand idea. She would help people without expecting anything in return. Instead, she would ask them to continue her act of kindness by helping someone else and ask that person to do the same.

A senior data scientist working at a local hospital, obsessed with spotting patterns, had just written out her latest findings about the flu season and was pushing her glasses up her nose as an oncoming gust of gray wind swept up her research and scattered it all over the road.

Lucy scrambled back and forth trying to hold on to her precious work as the papers slipped through her fingers. That's when she saw another pair of capable hands join in. At first, she questioned: Who is this? Why are they stealing my research?

In the span of those two thoughts, her papers had flown another meter away, and she brought her focus back to them as she leaped and reached to catch as many flailing sheets as she could. When, finally, there were no more to be caught, the stranger passed the papers to her. She then reassessed the stranger who had unbelievably intervened and joined the mission to save her research ... seemingly to lend a helping hand?

She narrowed her eyes and asked, "What do you demand in return? I don't want to be in debt to a stranger."

The stranger held out a hand and smiled kindly. "Penny, pleasure to make your acquaintance. You don't have to pay me back. Instead, I would like you to extend a helping hand, to no benefit of your own, and ask those that you helped to continue this chain of kindness onwards."

Penny smiled and carried on, leaving a confused Lucy to arrange her thoughts.

One day the following week, the perfect opportunity presented itself to her on a silver platter. Her research about flu season was swirling around her head as she leaned on the bus window during her journey home. Even now, it bamboozled her! During flu season, it turns out, for most other minor ailments like the common headache or stomach ache, going to the hospital is more dangerous than not going: you are exposed to the influenza virus, thereby greatly increasing your chances of catching it.

That's when someone interrupted her thoughts, "Hey, Lucy! Fancy catching you out here on the bus."

It was her nosy neighbor, the chatty one she usually avoided.

"You know, I feel like I never see you around our neighborhood, but what are the odds of meeting you on the way to the hospital with a terrible headache."

Lucy's eyes widened, "Did you say you were going to the hospital for a headache, right in the middle of flu season?"

"That's right, hmmm."

"Let me stop you right there in your tracks. Do you know what research I've been working on for the last quarter?"

And so, she carefully and logically explained how going to the hospital now was actually more dangerous than just staying home, backing her claims with solid evidence.

Just after getting off the bus, the neighbor asked, "But why're you telling me all of this Lucy? What do you want from me?"

Itching to tell her to stop playing her music so loudly after 11:00 pm, she bit her tongue and remembered Penny's sincere request "All that I ask is that you help someone else and tell them to do the same, to no benefit of themselves."

They hugged and parted ways to their respective houses, both buzzing after their unusual encounter.

Little did they know that their entire conversation had been filmed by an aspiring youtuber that sat behind them, intrigued that a person going to the hospital for a mild illness may actually be worse off without considering the risk of being exposed to the flu.

As soon as he put it up under the title of "Hospitals – zones of cure or death?", the video started blowing up, exponentially increasing in number of viewers every second. By the end of the weekend, the video went viral and even got on the morning news as everyone was struck with Lucy's revelation: "Don't go to the hospital during flu season".

Lucy's recently released flu season study shot up in popularity. Much to the dismay of local hospitals, people were very reluctant to visit during the entire season. In fact, even in the cases of serious symptoms of conditions such as kidney failure, cardiac arrest, or cancer, people overlooked the dangers in order to avoid the flu — a classic exhibition of **loss aversion bias**.

This meant a great loss in projected profits for the hospitals, and a need to downsize their operations, something which they would later come to regret.

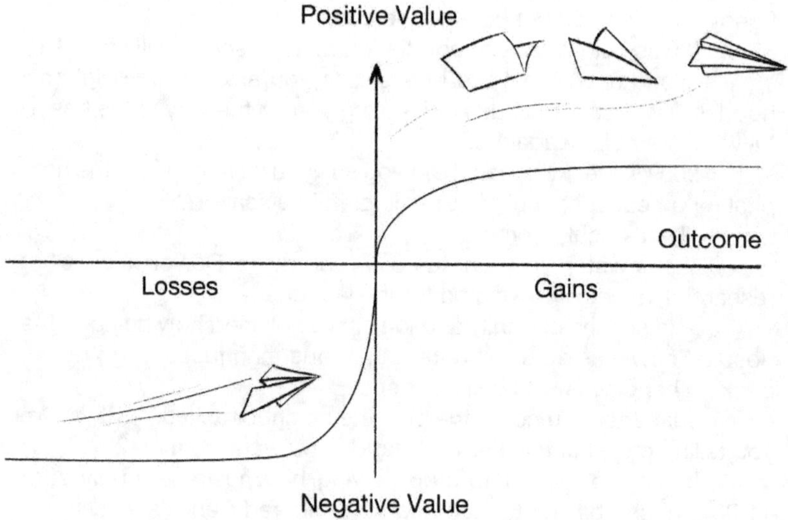

Further down the "kindness chain" was a strapping young chap named Griffin. He was an aspiring journalist, conducting interviews to furnish materials for his article "Life after a heart condition". He had conducted some research about food and nutrition for patients with a heart condition, and during one interview, he realized it was just the chance for him to perform his kind act.

"Listen, mate," he said, glancing around at the packets of polished white rice, "You just had a heart attack. Why don't you try reducing your intake of polished white rice?"

"Absolutely not!" was the unexpectedly stubborn response. "I don't think you appreciate that rice is fundamental in my food culture and is the backbone of most meals."

Griffin tried convincing the middle-aged man, but he held steadfast. That's when Griffin's natural storytelling instincts kicked in and he started hatching a plan, leaning on a white lie to support his "selfless kind act" and influence his audience.

"It would at least be healthier for your heart if you consider switching to a different brand of rice, Sona. They are cleaner in their process and proven to be better for hearts," he boldly declared, hoping his lie was convincing.

He justified this deception to himself as a kind act; what he overlooked was that Sona brand rice was more expensive, and this meant that this low-income middle-aged man who was just making ends meet between his minimum wage job and expensive medication would have to cut down on his rice consumption to be able to afford each meal with this more expensive brand.

Pleased with himself when the man agreed, Griffin bid him adieu till his follow-up interview the next month, hoping to explain his lie and also that since Griffin helped him out, he too would have to help another person without expecting anything in return.

With an extra spring in his step, Griffin arrived at the man's house one month later, eager to hear how he had adapted to his new expensive but reduced rice regime.

However, when he entered the house, he was surprised to see tons of Sona rice packets and hear that the man had in fact increased the amount of rice he ate!

Digging deeper, as any good journalist would, Griffin found out the counterintuitive reason.

"Originally, I had 100g of rice and 100g of veggies for each meal. Veggies are significantly costlier than rice. Constrained by the same budget to fill my stomach, I have to eat more of the expensive rice and substantially less of the even more expensive veggies."

Thus, an increase in the price of rice counterintuitively led to an increase in the consumption of rice. The kindness had backfired! Perhaps because it was seeded with a lie.

*Staple foods like rice tend to be a Giffen good for people in a low-income range, and therefore, as the price rises, the consumption also rises – the opposite of the expected effect. A **Giffen good** is a low-income, non-luxury product that defies standard economic and consumer demand theory. Demand for Giffen goods rises when the price rises and falls when the price falls.*

A season later, the kindness chain was still going strong and reached Camus, the enterprising owner of a small grocery store. He was going about business as usual when a phone call inspired him to carry out his kind act.

"Hello sir, is this Camus' Crib? The grocery shop off 15th Street and 4th Avenue?" the voice on the phone asked.

"That would be us! How can we help you?" Camus chirped.

"I've been driving for 6 hours and will be home in 50 minutes, right down the road from you. I have two kids in the car and no food at home since we are returning from a one week trip. I was hoping to stop by before your closing time and replenish my fridge, but alas, the little one threw a tantrum in the last service station that took quite a while to sort out. I'm a regular at your shop, the woman with the two kids and pink hat? Please, can you keep your shop open for just one extra hour? Don't worry, I will duly compensate you for your troubles," she begged, her voice ringing with desperation.

Now on a normal day, Camus would hang up straight away, paying no heed to this request, for never before had he stayed open a single minute past his 6 pm closing time. Having young kids of his own and feeling sympathetic to this woman, Camus decided it was time for his grand kind gesture, and he would surprise the woman by taking nothing in return.

This kind gesture, he was shortly to find out, would be a lot more effort than he had originally imagined. An hour passed, and another half an hour, and the woman did not turn up. Soon, she phoned almost in tears. There had been an accident on the highway that created a massive traffic jam, slowing down their journey greatly.

Camus sighed and rolled his eyes as he promised the woman that he'd remain open and await her arrival. Gosh! Who knew kindness could be such a drain on resources?

Eventually, instead of the original 6:00 pm, she arrived at 8:30 pm and made her purchases in a rush with a round of thanks. Just as she was reaching into her purse to give a generous tip, Camus reluctantly stopped her and passed on the kindness maxim, that she too would have to extend a kind gesture to someone else without expecting anything in return.

Finally, he managed to close shop at 9:00 pm and walked home, tired after an extra long day. What was odd though was how eventful it was. He thought being open the extra hours between 6–9 pm would be uneventful and dry, but unexpectedly, his business in those three hours almost matched the cumulative business from the entire rest of the day. As he walked home, much later than usual, he noticed his shop was the last to turn off his lights, as in this quaint neighborhood, everything closed fairly early.

Mulling over these thoughts overnight, Camus decided to try something new. What if he was the only shop to stay open late? As customers realized that, would all the business in those evening hours come to him? Surely it would be worth the added employee and overhead costs.

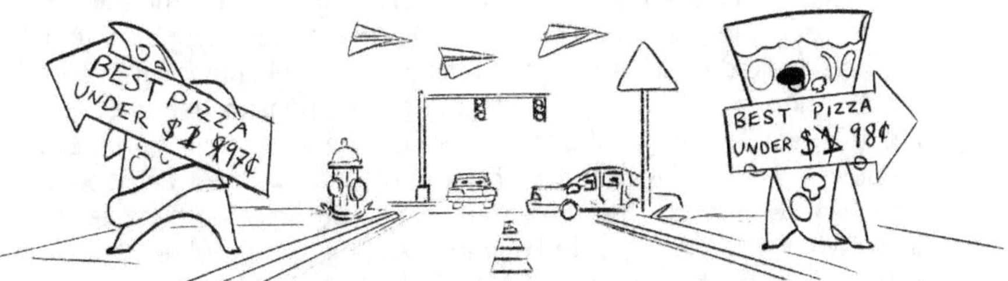

Much to Camus' chagrin, being a trailblazer doesn't last long when everyone sees your unbridled success. Soon, the town was abuzz about Camus' Crib's late hours, and this became quite a novelty. Seeing this, the other grocery shops grew nervous and realized that, to stay relevant, they had to compete and match up to these new targets. Within a few months, at least four grocery shops in the nearby area had extended their opening hours from 6:00 pm to 9:00 pm and were paying increased amounts in wages and overhead costs.

However, since now so many shops were open at night,

Camus wasn't seeing his previous profits. In fact, none of the shops were seeing a great increase in profits, since the customers were split between all of them. Additionally, since grocery stores mostly sold **inelastically demanded goods**, the net sales averaged out to be the same – only with higher wages and overhead costs!

This meant a net decrease in profits over time for all the shops involved. After Camus lost his advantage, it became worse for everyone, and no one can switch back since they will lose out. This is a clear-cut example of the **paradox of competition,** which in economics refers to a situation where measures, which offer a competitive advantage to an individual economic entity, lead to nullification of advantage if all others behave in the same way. In some cases, the final state is even more disadvantageous for everybody than before (for the totality as well as for the individual).

Just when this clique of shop owners thought they couldn't be any worse off, the kindness chain moved on...

"Clunk....tk...tk.tk." Jenna's thoughts came back to reality as a stone skittered across the sidewalk. She was in a dilemma and had unknowingly kicked the stone, and the non uniform gaps between the clunks snapped her out of her self-induced reverie.

She was worrying about a proposal that a private company had sent to her earlier. They had promised millions for her latest invention. She had worked on improving the efficiency of fuel engines for years with the hopes of passing it on to the government to gain recognition and quickly climb to the top of the national science institute. But the allure of a lot of money was hard to say no to. She was deep in thought when the chain of unnatural clunks brought her back to reality. "What was that repeated sound? Sounds like a broken chain in an engine?... Wait, a chain ..." And that's when she remembered the chain of kindness that should not break with her.

Thanks to the timely help of a stranger, she was alive today. She owed her life to the chain of kindness that had only one condition. Well, she was as arrogant as she was smart and declared "I will do much more than just pass on the chain to one person, I will pass it on to all of society."

The next day, the newspaper headlines read "Scientist Jenna improves the fuel efficiency to unprecedented standards and releases it to the government rather than private players. Now the environment and the common man's life will drastically improve."

The day that the newspapers were plastered with his wife Jenna's face and her unpremeditated act of kindness, Fallon was very confused at her behavior and concerned for her wellbeing. If she doesn't look out for herself, then who will?

He returned home after his day job as director of the state revenue services to find his wife waiting eagerly at the door, bouncing impatiently from foot to foot.

At first, he didn't want to hear any of it. It seemed almost as if Jenna had been brainwashed by some communist ideology, so illogical and selfless. The more and more she spoke, the more he was convinced he had to get her some help. She was blabbering on with bright eyes about "kindness" and "altruism" and "chain", and he was not going to play any part in it.

However, sensing that he was pushing her away, Jenna changed tact.

"As I said, you should increase the tax rates and invest all the extra money into homes for the homeless. The resulting social stability actually benefits you directly! The state will reduce spending on healthcare for the homeless, AND the newly homed can contribute to the GDP actively. Think about it! Imagine how much money you'll save for the state. And of course, the public is going to oppose the tax raise – but no worries, with the newly homed participating in the labor force, you will have access to a lot more taxable incomes and can therefore reduce the tax rates to even lower than before after a few seasons. Sounds to me like a win-win. I bet at the end of this campaign, you'll even get a bonus, if not a promotion," said Jenna with a proud smile on her face.

Fallon took a moment to let Jenna's words sink in.

"Curious plan. I'll think it over and run the numbers tomorrow. I suppose, there is no downside to it!"

Unbeknownst to the world, they were in for a ripple effect. In an oil-run economy, small changes lead to large side effects. When fuel efficiency increases, the cost of fuel for the same distance decreases; thus, it is more affordable for a whole new strata of the population. More people started using personal vehicles as compared to public transport, so buses which usually ran 20 trips with 50 passengers each time were now still running 20 trips but with only 20 passengers each way. Many changes like this were happening as the economy started shifting.

This counterintuitively led to significantly more fuel consumption and soon to scarcity all around. Shops like Camus' Crib, which were already stretched thin, now were faced with a fresh wave of problems as the supply chain broke down and essentials became rarer. To combat this scarcity, the government relaxed norms on production grade of fuel as they believed that the environmental positives of increased efficiency will outweigh the increased pollution of lower grade fuel.

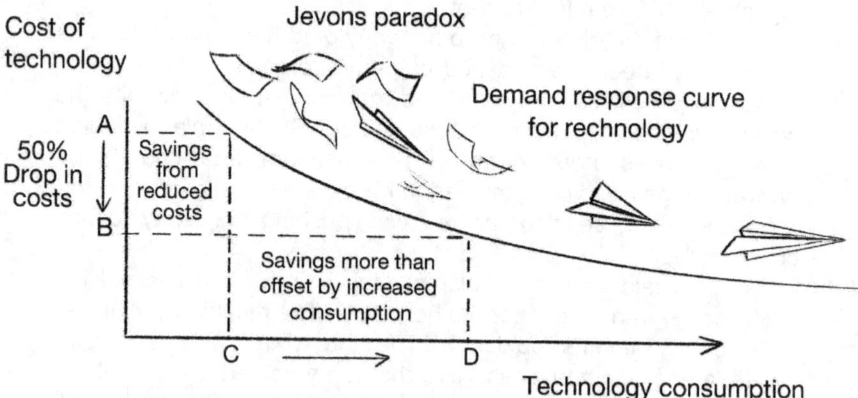

Consumption of tech more than doubles resulting in total costs being higher

*This situation gives rise to a classic example of **Jevons paradox**, one of the most common paradoxes in environmental economics. It states that, in the long term, an increase in efficiency in resource use will generate an increase in resource consumption rather than a decrease. This logic holds in the case of demand for fuel which is elastic in nature. Thus, as fuel efficiency increases its demand increase far outweighs its increase in efficiency.*

Soon, the lower grade polluting fuel coupled with increased usage started filling up the cities with smog. The fuel scarcity seemed to brighten things up for a bit before slumping back into reality. The only things that did better, although momentarily, were the hospitals. They had a sudden rush as many started contracting respiratory illnesses due to the sudden shift in air composition caused by the pollution. Alas, most of the hospitals had downsized operations, and this sudden influx was a nightmare leading to the collapse of medical infrastructure.

To top it all, Fallon's plan was in place, and the tax rates were also up. Counterintuitively, there was actually lower revenue from tax. This was due to a phenomenon called the **Laffer curve** – where the increased tax rate resulted in a portion of the population defaulting from the labor force and exploiting social welfare for their benefit. This is because they found that their best interests had changed with the tax increase, they would have to work more to get the same income, and they realized that actually it was more beneficial to not work at all.

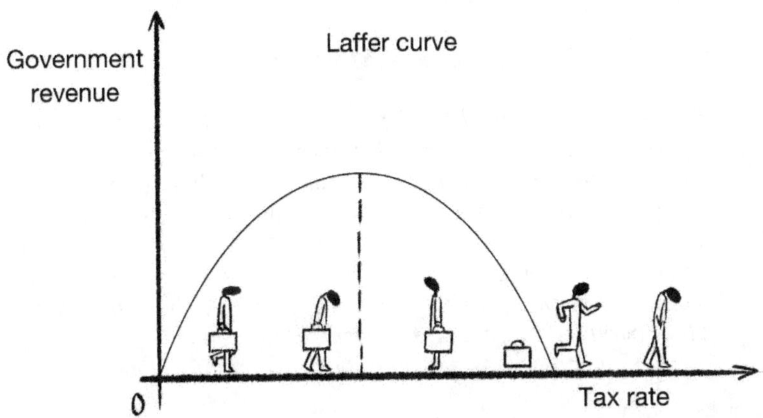

This put further pressure on the already sparse government finances, and thus large-scale oil or medical imports were also not feasible to mitigate the crisis.

During all of this, the government realized the political suicide associated with the problems of fossil fuels and tried to recover while heralding a greener future. The timeline for profitability for the oil corporations was narrowing. To make the best out of an approaching deadline, they had to optimize for short-term profit before their entire business went bust. This windfall of profit incited the fossil fuel corporations to speed up its usage further by investing in faster extraction and more efficient consumption – thus starting the cycle all over again.

Note to the Reader

Ethics, Kindness, and Altruism are at the end of the day economic strategies and not universal truths. Structures of society are geared towards certain strategies that survive, and those that lead to extinction are naturally pruned out.

Under certain circumstances, altruism is helpful and leads to a species' survival, making it a genetic trait. At the same time, other circumstances lead to it being detrimental to the extent of sometimes wiping out whole communities.

Well, who hasn't heard of the Trojan horse?

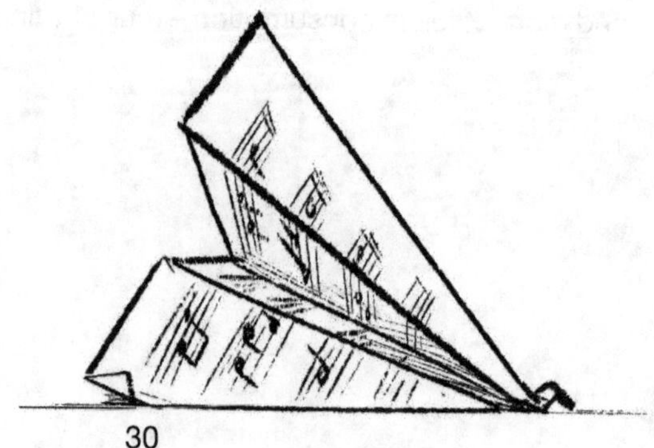

2.2
EXERCISES

1. Ultimatum game: needs two players.

   The game works as follows. Two participants are presented with an amount of money, say $10. The first participant
   makes an offer about how to split the $10 between each par-
   ticipant. The second participant can then accept the offer or
   decline the offer. If the offer is accepted, then the money is
   split as proposed by the first participant. If the offer is re-
   jected, neither participant receives anything.[2]

   Play the game a few times and observe what happens.
   Given this basic setup, how would rational economic theory
   predict that people behave? What do you think would happen if the amount was $10,000?

2. Three colleagues go paintballing, and the cost is $50.

   The three of them pay $20 each. The trainer hands back $10
   in change. The friends take $2 each and give $4 as tip back.
   They all paid $20 and got $2 back. $20 − $2 = $18. There
   were three of them: 3 × $18 = $54. If they paid $54 and the
   tip was $4 ($54 + $4 = $58), where did the extra two dollars
   go? ($60 − $2 = $58)

[2] Economic Puzzles (If You Don't Know Price Theory), Josh Hendrickson, March 11,2021. Economic-forces.xyz

3. The **St. Petersburg paradox** is a theoretical game, first proposed by Nicolas Bernoulli, in which you pretend that you are a player in a casino playing a special coin toss game. The casino starts with a guaranteed payout to you of $2. The game proceeds using a fair coin, tossed in succession until it flips a tails. After each flip where the coin is heads, the casino doubles the pot. So, if a tails appears right at the first toss, you get $2. If a tails does not arrive until the second toss, you win $4. If a tails arrives on the third toss, you win $8, and so on. The challenge is that you have to pay some amount of money to be allowed to play this game. If you were the player and were told to act completely rationally, considering only the expected payout and that the casino places no limits on the maximum payout, what is the maximum amount you should be willing to pay to play this game?

    Given this basic setup, how would rational economic theory predict that people behave? What do you think would happen if the amount was $10,000?

2.3

LOOK ME UP

Topic	Terms
Economics	Microeconomics, Macroeconomics, Supply and Demand, Elasticity, GDP, Inflation, Comparative Advantage
Utility	Behavioral Economics, Concave/Convex Utility, Marginal Utility, Diminishing Marginal Utility, Expected Utility, Budget Constraint, Income Effect, Consumer Preference, Pareto Efficiency
Risk and Loss	Risk, Uncertainty, Risk Aversion, Loss Aversion, Risk Loving, Prospect Theory, Insurance, Diversification, Adverse Selection
Game Theory	Strategy, Rationality, Utility, Altruism, Evolutionary Game Theory, Evolutionary Stable Strategy (ESS), Nash Equilibria, Schelling Points
Ideologies	Communism, Capitalism, Socialism
Propaganda	FUD, Dunning–Kruger Effect, Dihydrogen monoxide parody, Denial and Deception Framework
Other Relevant Paradoxes	Allais Paradox, Ellsberg Paradox, Paradox of Value, Paradox of Toil, Paradox of Thrift

2.4
NOTES

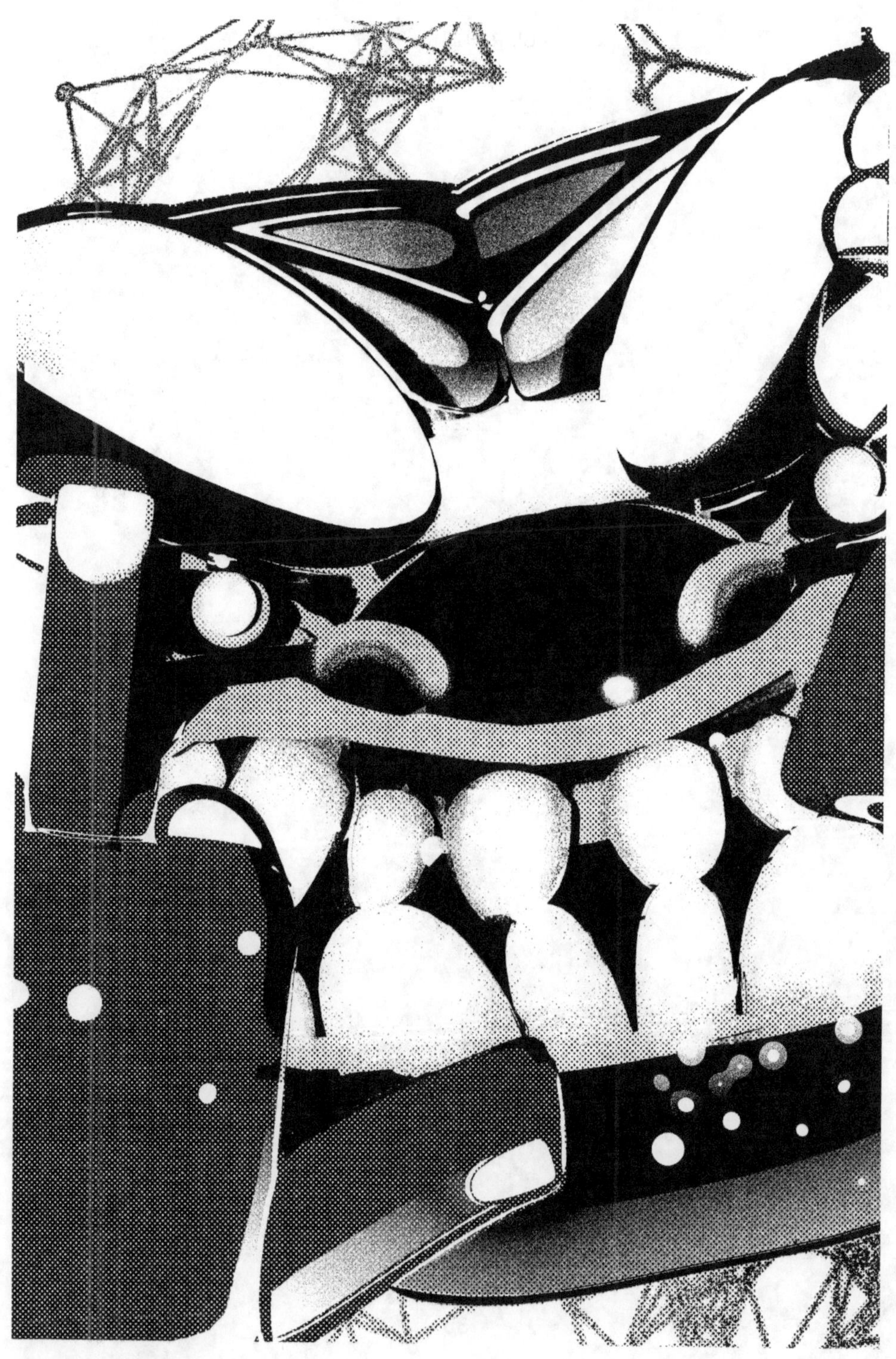

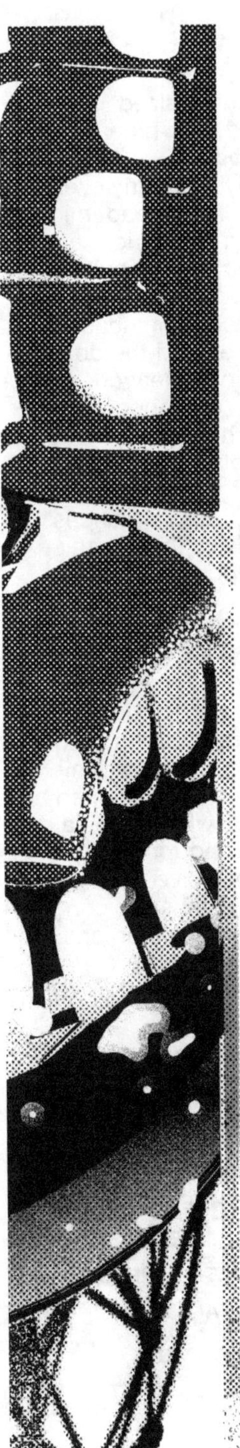

FAKEBOOK

A spoof HBR style three part article on the rise of Fakebook — an anti-social network.

The Man Who Only Knew Hate

The hate economy series

LET'S START AT THE BEGINNING. Park Xipperberg loved hating more than anything, so much so that he went and created the hate economy. Little did anyone know at that time, the power that hate held over the world. Fueled by divisive politics and an increasingly polarized citizenry, it was almost inevitable that Park invented the dislike button. What started with a simple button to express your political loathings soon spread like wildfire into every other sphere of influence, including brands, acquaintances, celebrities, public figures, foods, colors, pet peeves, philosophies, music, books, movies, and TV programs, to name a few.

Enter Frenemy, the social network built on enmity, where each dislike charts a graph of interconnected hates and you can easily find like-minded haters.

As the famous adage goes, "an enemy of an enemy is a friend." People soon found solace in their echo chambers. Whether it was dating better, hiring employees to fit the company culture, or finding like-minded friends, the network effects of Frenemy did it all. It fueled the flames of the previously untapped hate economy and grew to the behemoth it is today. With 84% of the world on it, not being on Frenemy, by design casts you as the universal enemy.

Traditional rubrics about 0 to 1 and **crossing the chasm** were dumbfounded by the unprecedented leaps Frenemy made. "0 to 1" essentially states that going from nothing (0) to something (1) is the greatest leap possible, even more so than going from 1 to n for instance, which is more about incrementally growing something that already exists — seemingly where the vast majority of human efforts lie. The infamous chasm between early adopters and mass usage was also easily overcome, propelled faster by hate than love.

TO GO FROM ZERO TO ONE IS TO CONJURE SOMETHING INTO EXISTENCE FROM THE DARK VOID OF OBLIVION. THIS IS THE ESSENCE OF TRUE INNOVATION.

— PETER THIEL

To explain in simpler terms, a loss is perceived asymmetrically to a gain. If you had a million dollars and I took away half of it, it would sting much more than the happiness you would receive if you got an additional half a million dollars. This can be easily observed when we see that a person who's broke gets much more utility from $10 than a millionaire gets from $10,000. Another analogy is that utility functions are concave and not linear, resulting in perceiving a loss as much worse than a similar-sized gain. So, Park believed that hate propels faster than love.

Surprisingly, Frenemy even broke previously held academic beliefs on the small world network. It was thought that any two people in the world were within six consecutive acquaintances of each other. If we extend the definition of small world network to include any type of connection, Frenemy showed how much more connected we are with like-minded hate and cut the chain length of intermediate acquaintances down to three. Even though this may have been possible with a 'love' network, none of them have a global scale of users since they don't exhibit network effects as well.

In this new world, brands, philosophies, and cuisines had a life of their own, grew organically out of hate, and reimagined the very notion of what a social network is. It was no longer just people; it was a massive network of echo chambers, shared consciousness, growing opinions, and, of course, a treasure trove of data.

WE GROW INTO A NATION OF HATE. WE HATE IN DEGREES, FIRST THOSE WHO CHALLENGE US, THEN THOSE THAT ARE DIFFERENT, THEN THOSE WE ARE TOLD ARE THREATS... HATE IS A TORCH THAT IGNITES EASILY, AND BURNS INDEFINITELY, NEEDING LITTLE TO FAN IT.

— SATHYA SARAN

THE WOMAN PLAYING GOD

The hate economy series

A trailblazing scientist with vision like no other. A dashing and charismatic single parent of three, an inspiration to us all. Plastered on the front of magazines, selected as one of the top 30 Under 30 scientists, youngest female recipient of the prestigious Fields Medal – the Nobel Prize of Mathematics – extensively profiled by the New Yorker, writer for Wired, and so much more; it was almost as if the men running these establishments felt guilty for never giving female scientists their due and chose this singular opportunity to make up for it all. Eliza Gomes has been the poster child of science for a good decade. That is, until now.

She plummeted from fame to notoriety when news of her odious rocket hit the world. The work of scientists over many decades led to one conclusion — humanity will be wiped out through nuclear war. QED.

The notion of **mutually assured destruction (MAD)** had kept full-blown nuclear war at bay so far, but each passing year grew more fraught with tensions. On behalf of the UN, Eliza biblically proposed an ark to be the savior of the human race.

In selecting those to be taken onboard the Spaceship of Survivors, Frenemy played an important role, for it served as the most holistic, comprehensive and complete data set of the world — by far more accurate than any census. Purportedly, Eliza's algorithm was designed to fairly and randomly select humans with some reasonable and necessary constraints and considerations, such as basic health factors, no outrageous criminal records, fairly preserving human heritage, ensuring procreation (would not be possible if all chosen were male) and no one over the age of 80. From this public declaration, hue and cry ensued all over the world.

I KNOW NOT WITH WHAT WEAPONS WORLD WAR III WILL BE FOUGHT, BUT WORLD WAR IV WILL BE FOUGHT WITH STICKS AND STONES.

– ALBERT EINSTEIN

Cue the intractable problem of choosing who deserves to be saved, for taking two-by-two of each race was still an intractable problem. Clearly, Noah was ahead of the game.

What does randomness here mean?

To randomly choose people within a sample set would mean to choose them unpredictably.

What does fair randomness here mean?

Fair randomness, in addition to being random, means that everyone has an equal chance of being chosen.

If we were to randomly choose without fairness, some

people could have a higher chance of being chosen than others. In the case of Noah's ark, if we randomly chose a couple from each species, it would be random but not fair. Animals belonging to a species with a smaller population (e.g., tigers) are more likely to be chosen than animals belonging to a more populous species (e.g., ants).

To ensure fairness, Eliza leveraged the vast Frenemy social presence to randomly select people (after crowd-sourcing the creation of dummy profiles for people who were not registered on Frenemy).

The world sees Eliza as a kind of Thanos (he killed half of all life at random, so that the rest can prosper), a harbinger of change for the greater good. The world is now waiting for the literal lottery of life, anxiously waiting for the result of this fateful exercise.

One of these drawings of random dots was made by a human and the other by a computer.

Can you tell which is which?

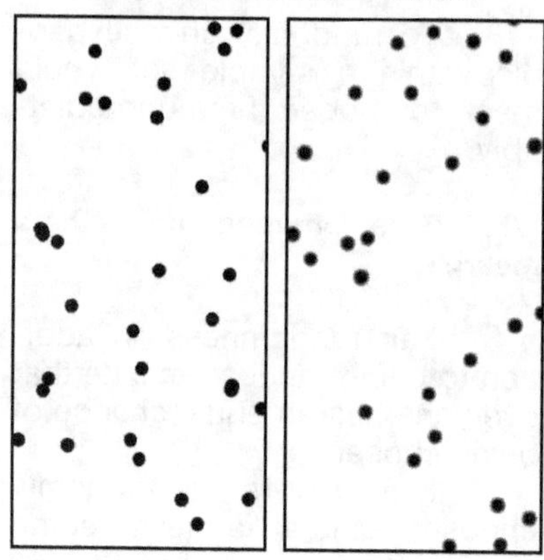

As they wait, all the armchair intellectuals come to life, bringing forth interesting open questions:

- What is the randomization algorithm/technique she used to achieve random selection?

- What should be the constraints to this random selection?

- What about pregnant women – are they treated as one or two? When is the fetus considered life?

- Male vs. female ratio – should it be equal? Even if the world's actual ratio is not? What about non-binary and trans folks? There is a small probability that everyone who is chosen is the same sex and can thus not procreate leading humanity to die either way.

- Should we have at least one of each race, culture, language? If we do, isn't it unfair to the larger sized demographics? But if we don't, we lose an integral part of living human heritage.

- What about separating families – picking a child without the parents would be detrimental to their development. So should families be chosen as a unit instead? Should people be given the choice to be a part of a bigger group and the probability adjusted accordingly?

And so it goes. The world waits, albeit impatiently.

THE CHILD WHO SAVED US ALL

The hate economy series

BREAKING NEWS

Never did we expect to break a scandal about the Spaceship of Saviours so soon. But should we have seen it coming?

The world sees Eliza Gomes as a harbinger of change for the greater good. With this controversial philosophy lies the truth we regretfully relay to you on this grim morning. She is much worse.

What originally seemed like an act of god turned out to be a calculating woman's ploy born out of deranged desires. The world was waiting for the literal lottery of life, anxiously anticipating their call to the Spaceship of Survivors.

What started out as a prank on his mother turned into a full blown investigation into deceit. Jules Assam, Eliza's firstborn who turned 18 this year, is perhaps the youngest and most unexpected whistleblower to date. When he breached his mother's work system, he expected to find nothing but boring calculations, constraints, and computing algorithms to support the random selection of humans from the Frenemy database. Instead, Jules found access to the entire Frenemy social network hate graph.

This piqued his curiosity. Why on earth would you need to access the entire graph and not just the nodes (which represented the profiles of users) when doing the random selection? This seemed too basic a mistake for Frenemy or, for that matter, anyone worth two cents of their computer science degree to make, a serious infringement of privacy and the public's trust.

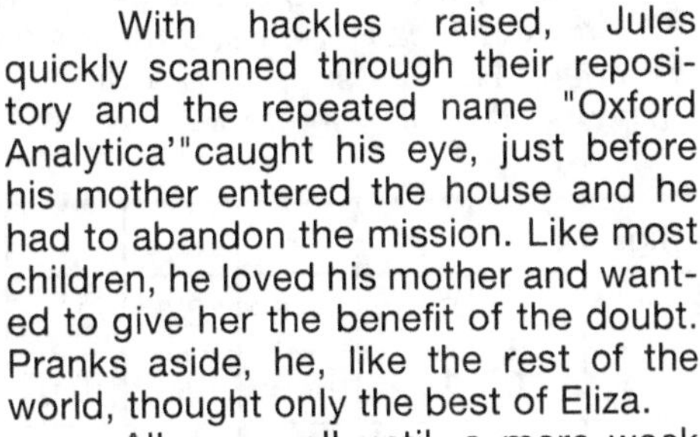

With hackles raised, Jules quickly scanned through their repository and the repeated name "Oxford Analytica'"caught his eye, just before his mother entered the house and he had to abandon the mission. Like most children, he loved his mother and wanted to give her the benefit of the doubt. Pranks aside, he, like the rest of the world, thought only the best of Eliza.

All was well until, a mere week

later, Oxford Analytica published an "independent" audit of the random selection process commissioned by the UN. The news caused Jules considerable distress. He felt that he was witnessing a conflict of interest at the very least, treason at the worst.

Jules, a novice in computer science, had a head-start in his investigation. Perhaps experts would've come to the same conclusion given time, but Jules knew what he was looking for. Using some basic techniques of fake data detection that he found online, he was alarmed to find further evidence of his suspicion. The report seemed to be almost entirely fabricated. And his mother's signature was plastered all over it.

Jules approached experts and news organizations, desperately trying to draw attention to his insights. It seems we are the first to verify his results. Eliza did NOT randomly select people, but leveraged the entire Frenemy hate network to optimize and find a clique of people with least internal conflict. In other words, her algorithm found groups of people with common hates, and only a few conflicting hates, so as to minimize disputes on the Spaceship of Saviours and beyond. And she consistently lied about it.

Let that sink in.

"The woman playing God" is right, for she decided for all of us.

For those that are technically inclined, we will walk through the techniques Jules employed to support his accusations. His techniques were rudimentary, but effective. We verified Jules' conclusion by enlisting experts to confirm his beliefs before publishing this article.

AN ERROR DOES NOT BECOME TRUTH BY REASON OF MULTIPLIED PROPAGATION, NOR DOES THE TRUTH BECOME ERROR BECAUSE NOBODY WILL SEE IT.

— MAHATMA GANDHI

To uncover the truth, Jules turned to a fundamental fake data detection technique, namely: **Benford's law**.

Jules took all the statistics and numerical data in the hefty Oxford Analytica Report and ran a simple frequency analysis on the numbers. He was supposed to find a frequency that looked something like the following table.

Digit	0	1	2	3	4	5	6	7	8	9
1st	-	30.1%	17.6%	12.5%	9.7%	7.9%	6.7%	5.8%	5.1%	4.6%
2nd	12.0%	11.4%	10.9%	10.4%	10.0%	9.7%	9.3%	9.0%	8.8%	8.5%
3rd	10.2%	10.1%	10.1%	10.1%	10.0%	10.0%	9.9%	9.9%	9.9%	9.8%

In large datasets, "1" has a little over 30% chance of being the first digit compared to a mere 4.6% for "9". But as we go down the digits of significance, the distribution tends to become more uniform.

Jules' finding did not match such a statistical distribution at all. This disparity does not directly imply fake data but is a great reason to investigate further. Jules strongly felt an urge to do the right thing and contacted multiple news agencies emphatically stating that there could have been fraud and it was worth double checking the sources of the Oxford Analytica Report. He also stated that he has no solid evidence apart from what he saw: that his mother had access to the Frenemy graph and had the Oxford Analytica name peppered over her repository. Failing the Benford's law test was the last straw that pushed him over the edge of suspicion to accusation.

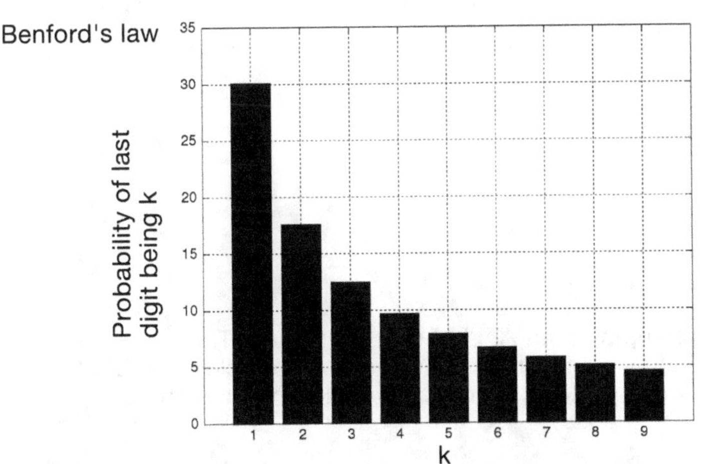

We dug further and discovered the shocking scandal behind all of this. The Oxford Analytica Report is a sham and authorities have discovered that it was an attempt at an elaborate coverup to hide the fact that Eliza was NOT randomly selecting people. More in-depth investigations are to follow, and one thing is for sure – that Eliza Gomes is going to get a little more notorious.

We only pray the trust in honest scientific research and institutions will not suffer too much from this incident. In the meantime, we will continue to be your most reliable source on the latest developments on this issue while upholding our standards of journalistic integrity and accuracy. You have the right to be informed as this is literally a matter of life and death for you.

More on this real soon. Watch this space.

3.2
EXERCISES

1. Select a dataset of your choice. Create a convincing report on the insights gathered from the dataset. Create a convincing report with fake data. Exchange with a classmate and ask them to figure out which one is fake and which one is real.

2. Use prompts on an AI LLM chatbot (chatGPT, Bard, BingAI) to conduct fake data analysis on a research article using **Zipf's law** and **Benford's law** and check for uniform randomness.

3. Use the **Hamming code**, a popular linear error detection and correction method. This exercise will take you through the steps of creating a Hamming (7,4) code, which takes 4 bits of data and adds 3 parity bits for error detection and correction.

Exercise: Suppose you want to send the 4-bit binary data 1101.

a. Calculate the **parity bits**: Using the Hamming (7,4) code, calculate the parity bits for this data. Remember that Hamming code uses parity bits at positions that are powers of 2 (1, 2, 4, etc.), and the remaining positions are used for the data bits. The parity bits are calculated as follows:

   i. P1 checks parity for all bits in positions whose binary representation includes a 1 in the rightmost position (1, 3, 5, 7, etc.).

   ii. P2 checks parity for all bits in positions whose binary representation includes a 1 in the second position from the right (2, 3, 6, 7, etc.).

   iii. P3 checks parity for all bits in positions whose binary representation includes a 1 in the third position from the right (4–7).
   Note: Parity can either be even or odd. For this exercise, let's use even parity, which means the total number of 1s in the bits checked must be even.

b. Encode the Data: Once you have calculated the parity bits, encode the original 4-bit data with the parity bits in their respective positions.

c. Simulate an Error: Now, imagine that an error occurs while transmitting one of the bits (you can choose a random bit to "flip").

d. Detect and Correct the Error: Use the Hamming code's error detection and correction to identify the erroneous bit and correct it. The position of the erroneous bit is found by checking the parity bits. If a parity bit is incorrect, its position number is used in a binary sum to determine the bit that caused the error.

e. Verify the Corrected Code: Finally, compare the corrected code with the original Hamming code to confirm the error correction.

4. Find out which of the following sequences are most likely fake generated data and the reason why. Feel free to use code, but it is possible to do it visually with some good observation skills. A bit of counting might also help in some cases.

   a.   11101110111111110111111111110111100100100111111101
         11110110100111110010110111111110011110011000011111

   b.   00011100101011101010110011001110010010101010010011
         10111001101010011000001101001111010011111110101010

   c.   01000000001101000101001110100100111101010000 1100
         01010010010011110001011001110011011011000010 0101

   d.   01010101010101010101010101010101010101010101010101
         01010101010101010101010101010101010101010101010101

   e.   00101101010100011110110010100101111110000101001
         01101010001001010101100010010101101000101110110

   f.   00100010110101100110110100100110110101100110 1010
         11010110011001100010001000100010001001100010 1101

   g.   01101111011100010000011101011011001010111001111
         11100111111101010000100011101000001101111000000

   h.   100111111001100110011010100111111111111111110000
         101000001001100110011111100110010000101000000000

   i.   10101010010001001001010010010100101010101010010100
         01001010101001001010100110101001101010101001000100

   j.   10111011101111101101010010111101111011101111100100
         11101011101111001000100010011101011101111011110

   k.   01101110000010010001101101011101011000010111 0110
         00100001011111011100000001100001100110010100011

   l.   01110001001010111011001001010111011110101101 01010
         10101010111001010010101101110101010100111011 0111

   m.   01011001101100000111011010100011010110111111 10101
         11000000010110101001001001010100110011010010 1010

   n.   10101101010101010101101011010101110011100111 0101
         01101011101111011101011101100100101010010011 0110

3.3
LOOK ME UP

Topic	Terms
Graph Theory	Binary Relation (Irreflexive, Symmetric, Non-Transitive), Hyper Graphs, Simplicial Complex, Adjacency Matrix, Graph Laplacian
Social Networks	Page Rank, Stationarity, Dirichlet Sum of a Graph, Eigen Vector, Spectral Properties of a Graph, Random Surfer (with Teleportation), Clique, Club and Clan, Community
Other Relevant Paradoxes	Small World Phenomenon, Strength of Weak Links, Granovetter's Paradox, Friendship Paradox

3.4
NOTES

DEADLY SERIOUS PARTY

A David Attenborough-style documentary on the Birds of Utopia debating voting systems.

SIX THOUSAND MILES off the mainland lies an archipelago: a world unto itself, forged of lava and isolated for thousands of years. The islands are known for their famously fearless wildlife and as a source of inspiration for Darwin's theory of evolution. And that's just part of the story.

Due to a lack of predators, abundant natural resources gave rise to armchair intellectuals, otherwise known as the Birds of Utopia. In this fertile land, a cornucopia of delights with no predators or sudden violent deaths and lots of time to pass, these birds did what any rational species would do to pass their time – they discussed the merits of voting systems. Second in marvel only to the mating season, the **anarcho-syndicalist plebiscite** has arrived with the blooming of the notoriously aromatic Rafflesia this year.

The ritualistic display began with a flamboyant dance by the majestic, moonwalking Manakin to officially open the floor for debate. As the esteemed Chair took her place, a cacophony erupted to welcome her.

It was almost as if this was their coliseum, and the speakers were gladiators. With a zoom, pip, and a stately thud, the dust cloud slowly settled to reveal the anarcho-imperial triad, born and raised on a literary diet so extreme, only few survive. Single-Shot Sicklebill, Preference Pelican, and Nay-Sayer Nighthawk were some of the finest of their generation and the tension in the air was crisp in anticipation of a riotous debate.

Using vibrational conversion technology juxtaposed on dynamic neural fMRI scans, we were able to translate the Grate De-bait into a humble human tongue with the aid of **Chomsky's linguistic theory of universal generative grammar**.

Let us jump right in.

– Recording begins –

Chair: We are flocked here today to witness this year's annual Grate De-bait. The annual debate was born due to the atrocious rates of narcissism, armchair debating, and infanticide. For many years now, we have agreed that our anarcho-syndicalist commune wants, nay, needs a revolution! However, since we can't seem to agree on anything else, here we are, stuck in this loop. Anyway, let the debate commence!

Today's topic was brought forward by Clueless Cuckoo. Due to inflation and global climate change, we need a WingMinister to steer us through the murky waters. For this, however, we need to have a fair voting system. Let's vote for one.

Single-Shot Sicklebill, Preference Pelican, and Nay-Sayer Nighthawk, I welcome you to take the floor and pitch your solutions. Begin.

Single-Shot Sicklebill: Hey folks, I'm a simple bird with a simple message for you to take home with you today. Don't ruffle those feathers, support **first-past-the-post**! It's the simplest and easiest voting system; cast your vote for one candidate, and the candidate with the most votes is chosen.

First-past-the-post / Plurality voting

Nay-Sayer Nighthawk: Hold onto your wings. This nonsensical First Past the Post voting system can let the most hated candidate be chosen. Look at this projector and see the absurdity. No civilized society would adopt such a farce of democracy. Power up the BirdsEyeView Projector!

Disapproval Voting

Nay-Sayer Nighthawk: I instead propose the superior method of **disapproval voting**, where you vote for one candidate and vote against another. The one with the largest difference of positive votes over negative votes is chosen.

Preference Pelican: Wait, that is still susceptible to **tactical voting** – where you would vote for someone that is not your first choice just because you are sure the candidate you wanted can in no way be chosen due to lack of popularity. As if that wasn't enough, it also falls prey to **Duverger's law** – in the long run this method will result in a two candidate system. In effect, all the other parties will end up dying out, leading to bipartisanship. I don't want to be left fighting between the cocks and hens – two parties with similar ideologies thereby offering little choice to voters. I propose the **single transferable vote (STV)** method, perfect for an articulate and learned society like ours, where each person submits rankings for the candidates without worrying about tactical voting. Let me dive into how this would work.

Single transferable voting

Preference Pelican: Let us start with five candidates, A, B, C, D, and E. Once everyone's first preferences are in, it is fairly clear that nobody has received a clear majority. So, we remove the candidate with the least first preference and distribute those votes based on the voter's second preference. Since no clear majority has been achieved, we repeat this process till a candidate has crossed 50% of the votes. As you can clearly see, candidate B would not have been chosen in traditional voting but is the clear favorite when we consider rankings. This ensures no tactical voting as everyone will just vote for their true preference without worrying about whether their candidate even has a chance.

See! As simple and straight as the crow flies.

Single-Shot Sicklebill: The crow maybe, but certainly not as straight as the arrow flies. You forget **Arrow's impossibility theorem**, a social-choice paradox illustrating the impossibility of having an ideal voting structure. It states that a clear order of the collective preference of voters, i.e., a ranking, cannot be determined while adhering to mandatory principles of fair voting procedures.

Fair Voting Procedures

- If every voter prefers alternative X over alternative Y, then the group prefers X over Y.

- If every voter's preference between X and Y remains unchanged, then the group's preference between X and Y will also remain unchanged (even if voters' preferences between other pairs like X and Z, Y and Z, or Z and W change).

- There is no "dictator": no single voter possesses the power to always determine the group's preference.

Single-Shot Sicklebill: Here is a rudimentary example everyone including the pigeons can follow, known as **Condorcet's paradox**:

No matter who is chosen, at least two voters prefer some other candidate over the winner. So, if A won, Voter 2 and 3 both prefer C over A.

Condorcet's Paradox

Voter	First Preference	Second Preference	Third Preference
Voter 1	A	B	C
Voter 2	B	C	A
Voter 3	C	A	B

A Whimsical Warbler that has gotten confused with all the theory flies up and manically squawks

Whimsical Warbler: Why should we vote? I propose a dictatorship and nominate myself as dictator, since less birds have as much authority in this archipelago than I have.

*a confused pause ensues as the entire audience, speakers, and

*Chair slowly tried to parse that phrase**

Single-Shot Sicklebill: Did you dare to heckle me with an **Escher sentence**? Away with you, you scoundrel! I can't even...

raucous uproar from the audience

Chair: Order! Order! Yes, yes, that comparative illusion which seems to make sense at first but actually has no true well-formed meaning is exactly an Escher sentence. I won't stand for this tomfoolery in my session! Security, Albatross division, please airlift this nuisance out of my arena.

furious flapping sounds as the Whimsical Warbler is being arrested

Whimsical Warbler: My vote won't matter anyways – most likely it won't be the deciding vote, so why should I even vote? Due to **Down's paradox**, if you are all rational voters, then none of you should vote, since the cost of voting is greater than the expected benefits.....

the Chair gasps

Chair: Wait a second, you imbeciles! You are trying to vote to choose a voting system. What voting system are you going to use for this vote to choose the voting system? Did none of you fancy schmancy "anarcho-imperial triadists" register the recursive loop in this pointless exercise? I'm done with this cuckoo charade.
I hereby dismiss this year's Grate De-bait.

air trumpet sounds

— Recording ends —

As the sun sets, the uncertain and hurried closing ceremony commenced. As usual, the anarcho-syndicalist commune left the Grate De-bait unsure of what conclusion to draw and yet were already eagerly awaiting the next year's pointless Grate De-bait.

Perhaps, we have a lot to learn from these Birds of Utopia, but perhaps not from their debating techniques.

4.2
EXERCISES

50 Voters:

30 Hatched

20 Dotted

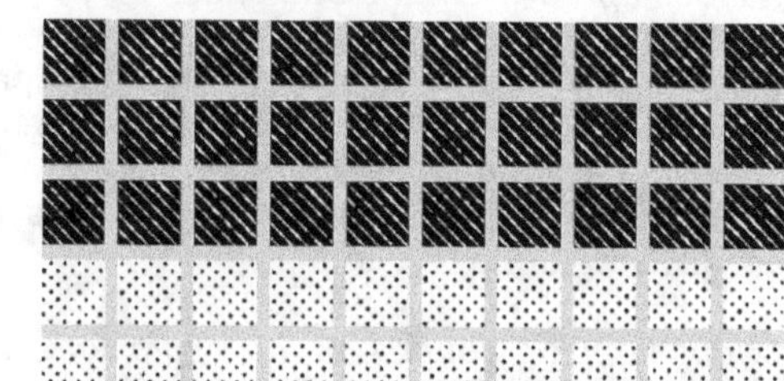

Districts Sorted by Pattern

3 Hatched Districts

2 Dotted Districts

Hatched Wins

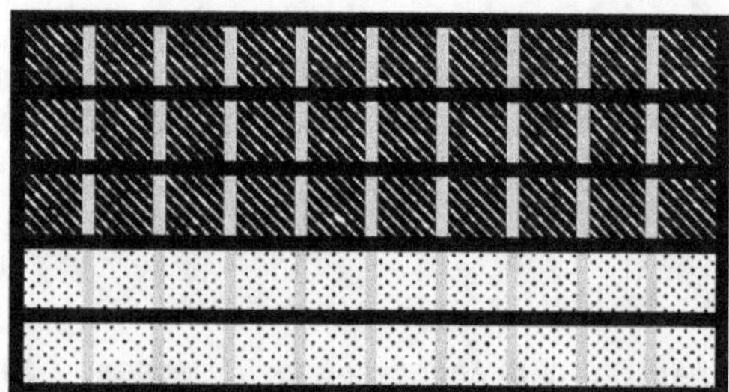

Compact Districts

5 Hatched Districts

No Dotted Districts

Hatched Wins

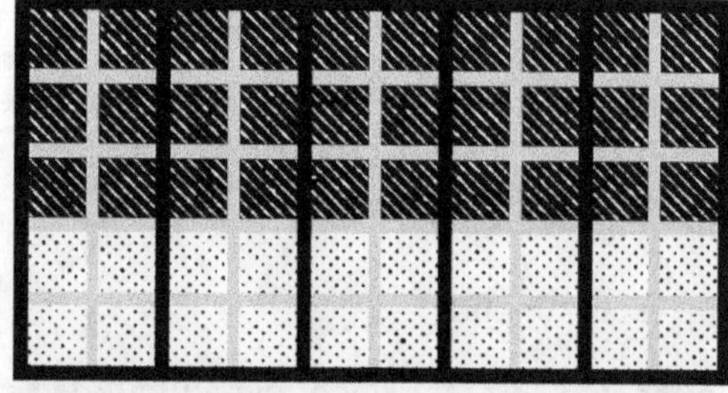

Gerrymandering

2 Hatched Districts

3 Dotted Districts

Dotted Wins

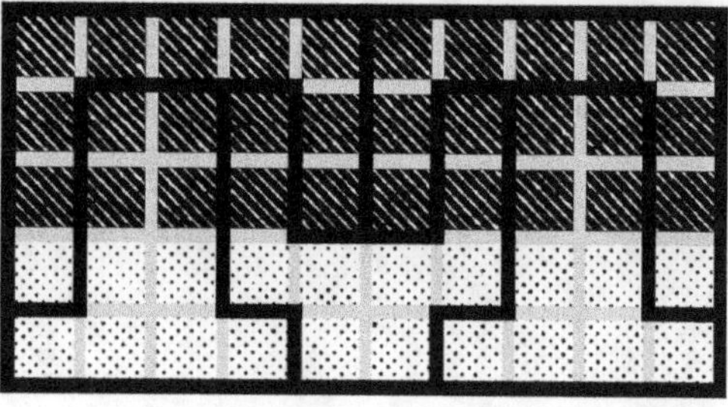

1. **Gerrymandering** is the art of redrawing voting districts to favor one party over the other. On the previous page are simple examples that show how drawing different districts leads to different winners.

   Your goal is to gerrymander the following areas to find a way to give each of the three tones the maximum possible seats individually.

   You may choose the size (>3) of districts as long as they are all the same size and are connected.

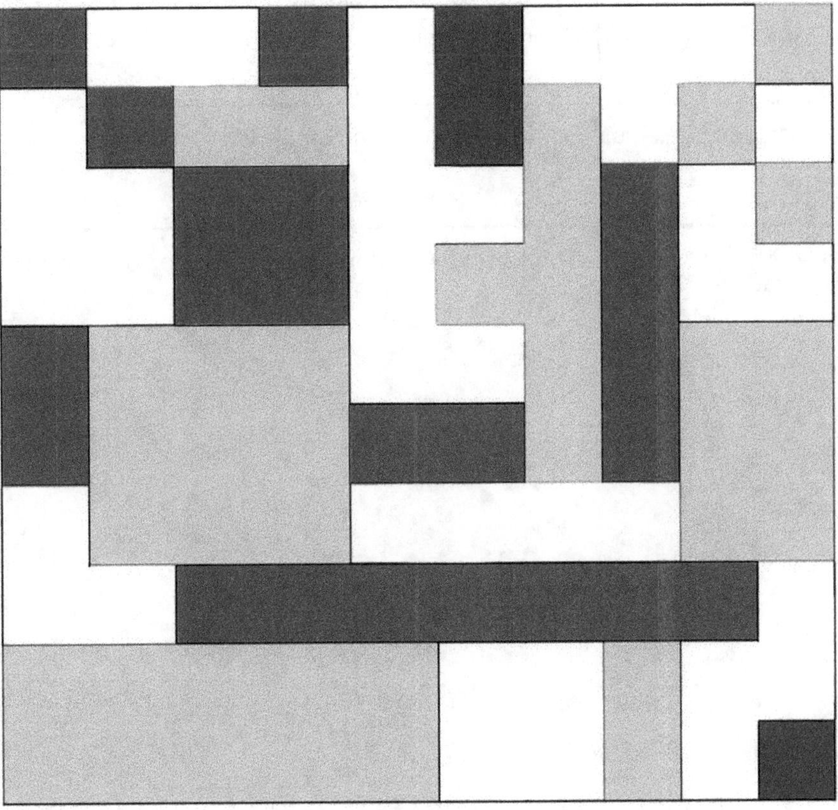

4.3
LOOK ME UP

Topic	Terms
Social Choice Theory	Voting, *Conditions for social welfare functions:* Unrestricted Domains, Social Ordering, Weak Pareto, Non-Dictatorship, Independence of Irrelevant Alternatives
Voting	Tactical Voting, Coalition, Ordinal vs. Cardinal Voting
Electoral Systems	Sortition, Liquid Democracy, Indirect Voting
Other Relevant Paradoxes	Condorcet's Paradox, Down's Paradox of Voting

4.4
NOTES

Schrödinger's cat

This is a poem, both healthy and strong
With rhyming couplets, an approach with prongs

About a curious case of a crawling cat
More perplexing than the antics of a frat

It got in a box, climbed in presumably
Along with a radioactive source, albeit stupidly

Both alive and dead
Can it even be fed?

Which begs the question: what will we see?
When we open the box, who will be?

Will the cat be poisoned and stiff with rigor mortis?
Or perhaps dreaming, fantasizing its next kiss?

And if we don't open the box,
what to assume?

Simultaneously being alive and dead?
Sounds like a curse, not a boon

This is also not a poem, but nonsense abound
For this quantum paradox, can never be found

Part 2

Pandemic Pandemonium

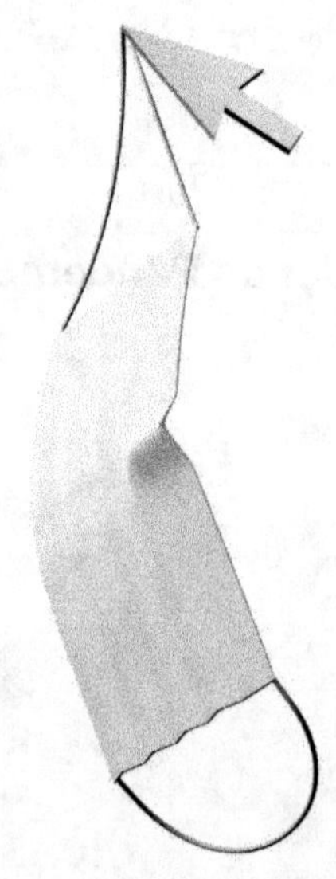

THE FOREVER VIRUS

A new editor tries to bring to light the bloopers that were published during the Covid-19 pandemic.

Hey, everyone! Please settle down. Let's get started. As you all know, I have recently joined the team as an editor-in-chief and love being onboard this rocketship! However, to set the tone of our session today, I was brought on to rectify our current editorial practice. Suffice to say the pandemic has transformed reporting as we know it, but we have to remember our basics. I've noticed a trend in the slip-ups in our editorial practice – specifically, when writers think it's okay to economically stretch the truth on scientific reporting.

Today is part one of a series that brings us back down to the basics. This initiative hopes to equip us to achieve the most accurate reporting we can during these troubling times and ensure our readers are well informed.

Let's jump straight in!

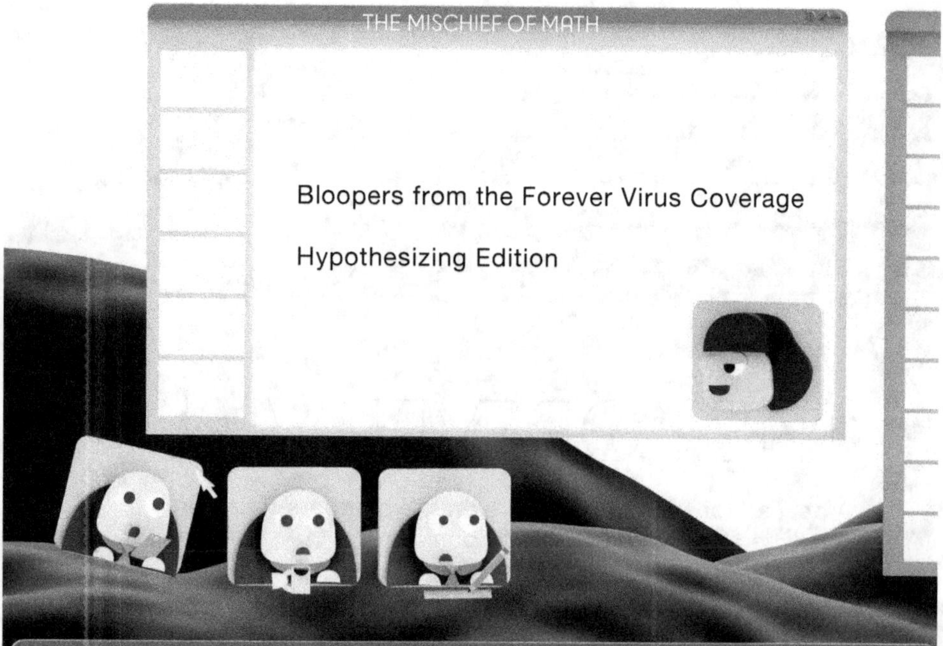

Bloopers from the Forever Virus Coverage

Hypothesizing Edition

The first part of this series will focus on hypothesizing. I will show a volley of different snippets from our paper reporting the pandemic over the last two years and call out the bullshit, if you'll pardon my language.

So, let's establish some ground rules. All scientific research starts with hypothesizing.

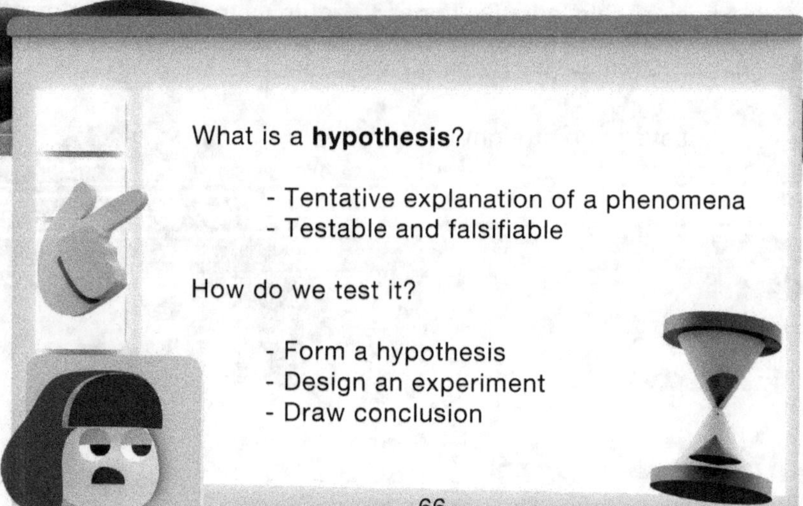

At the start of any research, a hypothesis is stated, and researchers endeavor to prove or disprove it. A hypothesis is a proposed explanation made on the basis of limited evidence as a starting point for further investigation. It must be testable and at the same time falsifiable. Don't worry about the jargon! A statement being testable or falsifiable means that it can be proven or disproven, respectively. A pithy example of a non-falsifiable statement is "God exists". This is why people consider the question of the existence of God not to be in the realm of science: How could you even test such a hypothesis?

Let's begin. A study analyzed eight countries, four of which had implemented lockdowns and four that hadn't. We drew up conclusions and held the front page for *Lockdowns Have No Effect*. Does anyone have an idea if there has been a mistake?

The Daily Circus

10.04.2020   Sunday

SHOCKING STUDY FINDS:

Lockdowns have no effect

A recent survey of eight countries has surprisingly shown that lockdowns have little to no effect on the rate of spread of the Covid-19 virus. It was observed that four countries that practiced lockdowns and four that didn't during the recent Omega variant have all shown similar R0 values.

This is just what I was worried about. It seems that the team is slowly getting into a dangerous echo chamber. Let's not indulge in **group-think**, folks.

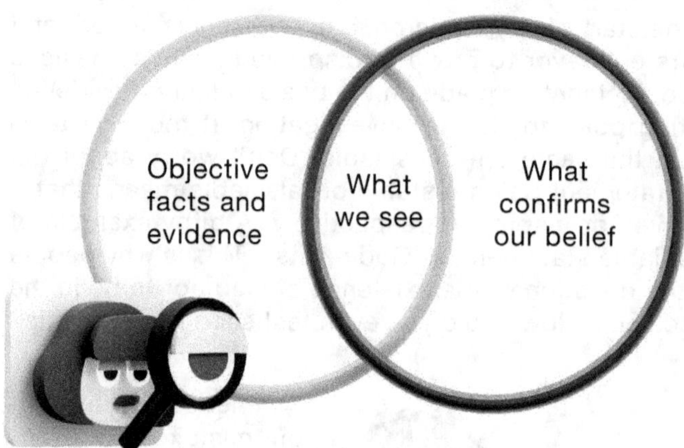

confirmation bias

We tend to focus on evidence that confirms our hypothesis and beliefs and unwittingly overlook other trends which may contradict them. This can be more dangerous than it sounds, especially when we journalists are trying to translate complex scientific results to simple English for our readers.

In the article, we had originally concluded that lockdowns have no effect solely based on the unpromising changes in the R0 value, but we didn't consider the demographics, vaccination status, climate, and cultural factors, which could have very well resulted in completely different assessments of lockdown. Spreading misinformation for a catchy headline is the worst thing a journalist can do. I know this was unintentional, but we gotta be extra mindful to uphold the integrity of our paper.

Don't hurt your head too much yet! We have a lot more to absorb today. Which brings us to another one of our headlines: *Why Obey Masking?*

Without getting into the politics of it all, any ideas what's gone wrong here? Besides the questionable syntax choices, that is.

It must be that the logic behind our argument is rather circular. We didn't give a proper reason and instead just told them that it was illegal to break the law.

This is formally called **circular reasoning**, and I'm surprised it only got caught here 'cause it's usually easier to spot this one. Anyways, I thought I'd also call out another variant of circular reasoning called

Curry's paradox

When the hypothesis is self-referential.

Since the claim, B, is arbitrary, any logic following this form of "If A, then B" where A is self-referential and True, will allow us to prove any claim B.

This next one is fun.
Should we trust experts?

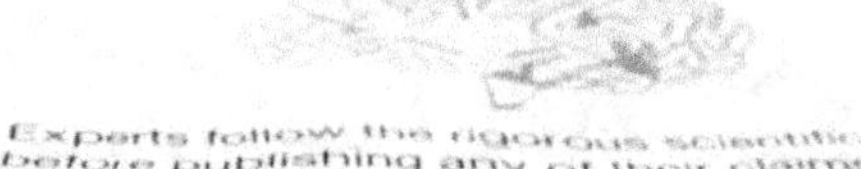

Experts follow the rigorous scientific method
before publishing any of their claims.
Therefore, we should trust in the academic
process and institutions and follow actual
studies rather than social media op-eds.

Here is something a bit controversial. In my opinion, an expert may be wrong, but on average, they get it right more than someone who is not an expert. The fact remains that the whole may sometimes be lesser than the sum of its parts.

Let's break it down to the acceptance criteria of a publication. For any research to gain traction or even cause a small ripple, it should be interesting. Of the hypotheses interesting enough to test, very few turn out to be true. Due to unavoidable errors in our measurement, there is a small chance that an incorrect hypothesis comes out as true (**false positive**) or a correct hypothesis comes out as false (**false negative**). So, false positives and false negatives do exist in research, and we should be cognizant of this. For example, a pregnancy test that indicates a woman is pregnant when she is not is a false positive, while a pregnancy test indicating no pregnancy for a pregnant woman is a false negative.

We have already filtered out the uninteresting hypotheses. Of the ones interesting enough to test, about 10% will be true. So, testing 1,000 hypotheses will likely result in only around 100 being true. The tests have a false positive rate of 5%.

False Negatives   Positives and False Positives

That means they produce 45 false positives, out of the 900 which are actually false. Out of the 100 which are true, they have a false negative rate of 20% so they confirm only 80 of the true hypotheses, producing 20 false negatives.

The researchers have no idea what is really true and what is really false. They just see an array of hypotheses of which 125 (80 true positives + 45 false positives) are true. With conservative estimates, we can already see that 45 of these are false. The negative results are much more reliable but unlikely to be published. Thus, consensus on trusting publications is a tricky subject, as we have put up a complex filter of acceptance.

Whew, that was a dense monologue from me! Everyone still with me? Don't worry! These slides are shared on the meeting invite, and you can go over it later.

So, to wrap up today's session, it is ok to make mistakes – that is part of being human, but not learning from them is essentially making the choice to be wrong. As a paper, I think it's time we turn over a new leaf and take this opportunity to refine our editorial practice. Scientific reporting, especially during a crisis, is of the utmost importance. It is our duty to keep the public calm and well-informed. I understand we need striking headlines to attract the reader's attention, but let's not compromise on quality and trust in our brand. From here on out, we can work together to raise the caliber of our output. The Daily Circus has a tremendous legacy, and now it is in our hands.

5.2
EXERCISES

1. Find an article in the last month which exhibits a paradox or bias.

2. Which of the following are True / False? Provide a reason why:

   a. The hypothesis is the first step of the scientific method.

   b. A hypothesis must be testable and unfalsifiable.

   c. The **grue paradox** is an example of a temporal assumption.

   d. **Confirmation bias** is when you highlight evidence that contradicts your beliefs.

   e. False positives and false negatives exist in research.

3. Find at least one example of confirmation bias exhibited in the following areas:

    a. By your professor in this course.

    b. By your university/college/school in any of its communications.

    c. By the authors of this book in this book.

5.3
LOOK ME UP

Topic	Terms
Hypothesis	Occam's Razor, Predicate, Falsifiability, Paradigm Shift, Not Even Wrong
MVP	Minimal Viable Experiment, Business Model Canvas, Lean Architecture, Zero to One,
Other Relevant Paradoxes	Grue Paradox, Goodman's Paradox

5.4
NOTES

6.1

ENDING THE FOREVER VIRUS

Selected entries from the
WHAT (World Health Awareness Trust) General's daily diary.

Dear Diary,

The sky, like my mood, is gray and overcast. It's been days since we declared COVID-19 a global pandemic, and yet people and governments alike are behaving ludicrously! Have citizens no faith in institutions and experts anymore? If so, my role is rendered defunct.

I am getting more and more concerned as I watch humanity tear itself apart. Today, I learnt of some horrific parties called "COVID-positive", positively celebrating COVID positivity. Apparently, true to their name, these unspeakable parties are hosted by individuals who are COVID-positive cases confirmed by the state. The infected host creates a large bowl of punch and spits into it, while all the party-goers drink their share. This foolhardy protest was supposed to show the world that the COVID craze is a hoax, no worse than the "common flu". So, on one hand, we have these naysayers, and on the other, we have the over-paranoid hoarders. When I went to the market today, literally all the tissues, sanitizers, soaps, gloves, and masks had been wiped clean off the shelves.

I hope that the paranoids realize that they are causing much more harm to themselves than good. By hoarding these essential supplies they are ensuring that others in their neighborhood lose access to them. The unprotected neighbors with no access to preventative resources are now much more susceptible to catching and spreading the virus, thereby increasing the risk for the entire neighborhood, including the very hoarders themselves. It seems counterintuitive, but if a hoarder shared their resources, they would actually be much safer than keeping them all.

"Sharing is caring" rings true once more!

This situation represents basic **zero-sum bias**, where a person believes that their gain has to be another's loss. Little do they realize that everyone can gain and lose together. The most surprising part is that even countries fail to realize this basic concept. COVID anywhere is a threat to humanity. So, essential medical supplies and personal protective equipment (PPE) should be adequately shared to reduce the risk of viral mutations. Ultimately, it is better for us to conquer COVID pandemic as a planet and not have a whack-a-mole of a game – individual territories heal, only to re-surge with a single super-spreading visitor arriving from another territory.

If you thought it was already tough to pass on common sense as an intergovernmental global organization, imagine what happens when advice gets mired in a whole lot of jargon. For example, today I heard in some lecture by the bio boys "Let's start with the hypothesis that Tregs play a role in the pathogenesis of COVID-19." T-rex what?! I suppose I'm going to have to familiarize myself with ALL the terms of virology, immunology, sociology, and epidemiology because it looks like that's going to be a shadow over me for the foreseeable future.

Sigh. I just hope people take this seriously really soon, or our mortality count statistic will snowball. Especially for the wise-and-elderly.

Yours Truly,
   The General of WHAT

Dear Diary,

Today's NEWS was a great shock. It read that:
"85% of the people who wear masks get COVID."

My first thought was that it was clickbait or fake news. But a more thorough read showed that the president of the world's largest economy had stated this misleading nonsense ON RECORD! Does this mean that wearing a mask can lead to COVID? A very confusing and contradictory message to be getting from the authorities, which had previously ordered mask mandates.

Obviously, I had to dig deeper and get to the bottom of it. I am ankle-deep in a pool of articles and reports. Ironically, my prosecution of the case led to the classic **prosecutor's fallacy**. The basis of the shocking statement comes from a widely conducted study that stated that out of those who got COVID, 85% stated that they had worn masks regularly. But the president misinterpreted it to mean 85% of the people who wear masks get COVID. It's almost as if wearing masks leads to COVID infection. How crazy is that?

It is like reporting "100% of those who got a job have a degree from NYU" instead of "100% of people who get a degree from NYU got a job." Absurd!

For this exact reason, I get anxious about every sentence I utter in fear of being misinterpreted. It's wondrous that the highly esteemed president had no reservations in announcing any sort of distortion of statistics, perhaps precisely because they intend to cause confusion and strife. How are we to do our jobs and encourage rational, sensible behavior when the people of the world are being assaulted from every direction with truths and falsehoods intermingled?

I really do fear for our peace of mind. If not, it will be the anxiety that kills us.

Yours Truly,
    The General of WHAT

Dear Diary,

My day began with a shock. I literally dropped my morning coffee when I heard a puffed-up reporter on TV report that smoking could be protective against COVID. How ludicrous! Any toddler with common sense could tell you smoking is terrible for the lungs.

The reporter went on to show how tests for COVID-19 on hospitalized patients showed lower rates of infection amongst smokers than non-smokers. Also, the proportion of smokers amongst those hospitalized with COVID-19 was less than the proportion of smokers in the populace at large. For instance, in China, 8% of those hospitalized with COVID-19 were smokers, when 26% of the general population smoke.

I had to dig deeper into the numbers and finally my team figured it out. A standard **Berkson's paradox** at play. :(

Let me break this down to make sure I got it:

=> Both smoking and being infected with COVID-19 are reasons for being admitted to the hospital.

=> Let's assume that patients are admitted to hospitals for symptoms due to only two causes: smoking and CO-VID-19.

=> Within this hospital population, most smokers will tend to show lower rates of COVID-19 as they were mostly hospitalized for exclusively smoking induced illness and not necessarily also COVID-19.

=> We see an emergent relationship between smoking and COVID-19 owing to sampling a selected population.

=> Thus, it feels like more smoking leads to less severe COVID and vice versa...

=> WHICH IS OBVIOUSLY INCORRECT!

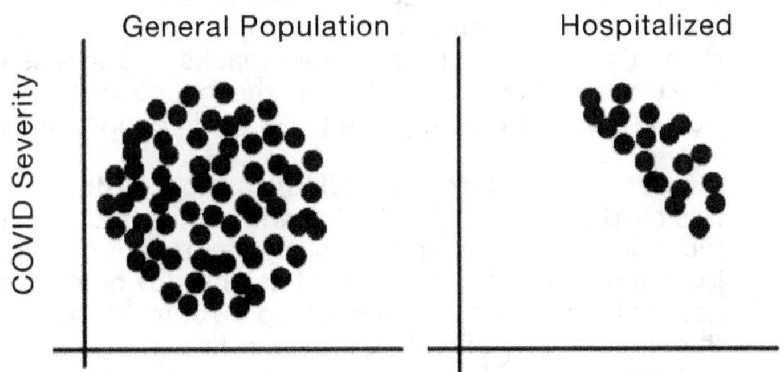

How are we gonna get over this damned virus if news agencies and scientists are basically conspiring against common sense?!

Yours Truly,
       The General of WHAT

Dear Diary,

"All of humanity's problems stem from man's inability to sit quietly in a room alone."

Regretfully, the time has come to fall into disrepute as an institution. We are already unpopular in the eyes of public opinion with our staunch professional recommendations, such as protective masks and incessant lockdowns. Now, if we reveal the success of lockdowns (or lack thereof), the people will be incensed.

I'm not saying there is no effect to the lockdown, but it is hardly worth it. The lockdown fatigue and general plummeting morale isn't even the main reason – if you look at the economic costs of this shutdown, it's been astonishing. Whilst conjuring up a combat plan for this virus out of thin air, we took a firm stance on the welfare of the citizenry – when there is so little known about the virus, the risk is great. However, maybe it was a little in haste...as the shopping malls, cinemas, travel, restaurants, and so many more have either suffered devastating blows or are altogether driven straight out of business because of unhealthy supply chains. For the folks in these industries, what does social responsibility and public welfare matter if they don't have bread and butter to feed their children?

Many governments even twisted our welfare pre-rogatives for political mileage. It led them through unwinnable elections and swayed public opinion in their favor. But now it is coming back to bite them. Serves them right for corrupting a massive health crisis with their dratted agendas.

Initially, lockdowns were implemented all around the world, but now that it is evident that they do little to help, it is apparently political suicide for governments to stop lockdowns as they cannot justify the resources and time lost to it already, thus admitting to their ineptitude and accept loss of face – exhibiting the **sunk cost bias**. The inertia is too great to stop this fallacious waste of time, resources and efforts.

So these draconian measures have to stop, for they benefit no one. It's clear it's not working.

Yours Truly,
    The General of WHAT

Dear Diary,

Finally some good news! At least in theory.

We discovered a wonderful concept called
Parrondo's paradox which states that:

"It is possible to combine two losing strategies to create
a winning strategy by playing a move from each losing
strategy alternatively."

When every strategy considered so far seemed to come
at the cost of either economic health or population health,
now we realize that there is a way to balance optimizing
both of them quite neatly – simply by alternating.

Losing strategy 1 = go into complete lockdown for all
ages, thereby safeguarding population health at the extreme cost of economic health

Losing strategy 2 = open up the economy completely and
let the population roam freely, thereby safeguarding economic health at the extreme cost of population health

If either of these strategies were followed on their own, it would be disastrous. But alternating between both for an optimal time period has proven remarkable results in our simulations! Of course, the time period before switching strategies would be different for each country, as their demographic, medical, and familial structural statistics differs greatly, e.g., Afghanistan typically has large families that live close by to each other while the Netherlands has very small families on average and live quite distantly – so the length of time they have to spend in lockdown before switching could be quite different.

If governments investigate the ideal time period to play each strategy before switching, it could very well lead to a viable and balanced solution. In fact, a researcher was saying that something curious called a "**Turing Labyrinth**" of herd immunity could be formed. In other words, alternating these losing strategies can stimulate viral spread in a pseudo-controlled manner, and the recovered and thus immune communities then form the walls of a labyrinth that the virus cannot travel through, thus separating the infected from the vulnerable. It's almost like we're barricading the virus in little pockets in a maze of immune people – reducing crazy spreading and isolating hotspots!

Interestingly, the simulations accounted only for population health and economic cost. We could have modified the model to include any other characteristic such as mental health or even environmental damage.

We are limited only by our imagination and ingenuity.

Yours Truly,
      The General of WHAT

Dear Diary,

Another day, another mess to clean up.

If I hear hydroxychloroquine ONE more time, I swear I'll hit someone. Bloody well time that the paper is rescinded before causing more confusion.

The researchers involved in the now very controversial

hydroxychloroquine paper stupidly had a six-day follow-up period as the inclusion criteria for the patients. In other words, people who died or became seriously ill during that period were not included – only those who survived were considered for the trial. Obviously, the patients who attended the trial were healthier and would show a markedly better performance than the original cohort of experimental subjects – which includes those who could not pass the six-day follow-up period. This led to incorrect causal inferences that hydroxychloroquine could be used to treat COVID, and voila, global misinformation spread, and chaos ensued!

This paradoxical situation is due to **survivorship bias**

– a logical error where a skewed sample selection process looks only at specific segments of the population that have survived through a course of time while discounting samples that have failed during the same period. This bias results in a skewed perspective that fails to account for the larger picture or complete reality.

It is easy for authority figures who wish to sway public opinion to buy into these lauded but faulty clinical trials – they can cite specific examples with positive outcomes to veil the grim reality.

But this is not the truth.

This entire episode once again gives voice to the old adage: Death prevented hospitalization, instead of hospitalization preventing death.

Yours Truly,
        The General of WHAT

Dear Diary,

With all the hullabaloo about vaccine trials, it seems that this vaccine race has devolved into a comical game. It is becoming a target game with each company touting better and better scores. Under the hood, the intricate details stopped mattering, and everything we see in the newspapers now starts to blur. So, the really important numbers are seriously getting lost in this pigsty of "Scientific Discovery".

This rant does not come unprovoked.

Today, whilst doing my due diligence on the AstraZeneca vaccine numbers, I stumbled upon an impending press release on my desk that had headlines that really made this pandemic response seem like a parody. How did such flawed scientific research even get so far?!

To break it down:
It was just **confirmation bias** in action; these guys moved the acceptance goalposts by giving a cohort a head start. Remember when it was proudly reported that AstraZeneca was the best COVID vaccine? Well, that conclusive headline turned out to be misleading as the study that was cited was not properly inspected.

Accidentally, a smaller preliminary dosage was administered to a Brazilian cohort, which seemingly coincidentally had a better performance later. Basically, the Brazilian cohort was administered a fraction of a dose before they entered the experiment to check for allergies. This happened well in advance and actually acted as a pre-booster, resulting in improved immunity and therefore a bump in the efficiency of the vaccine as a whole. In essence, they contaminated the trial before even beginning it.

It suffices to say that Astra-whatever is not entering my household ever.

Yours Truly,
    The General of WHAT

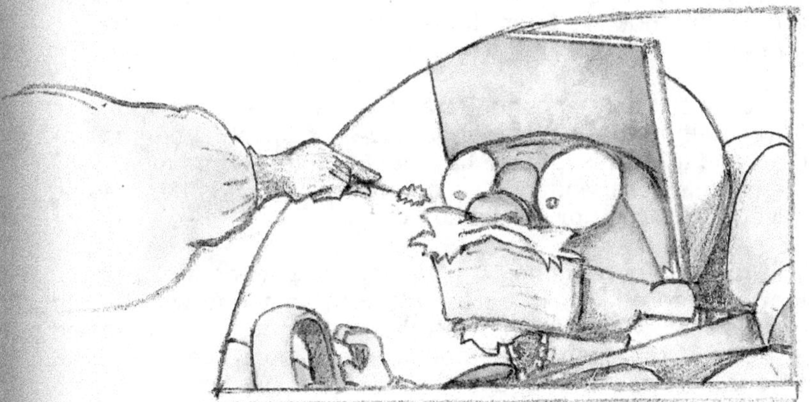

ENDING THE FOREVER VIRUS

Dear Diary,

Today, a couple of young Courant grads came to me with an interesting proposal. For the first time, they brought a counterintuitive graph theoretic concept called the **friendship paradox** into vaccination deployment strategy.

Traditionally, vaccinations are done in waves of urgency followed by necessity and then rolled out to the general public, i.e., older or at-risk patients with comorbidities, then front line and healthcare workers, and then the rest of us. But when it comes to us, vaccination is purely ad hoc and unstructured. So these younguns came in to improve the mass rollout through a novel method with very little extra computational resource and effort.

The crux of the Friendship Paradox is: an average person has fewer friends than their friends have. Intuitively, you can understand that a few people have lots of friends, and most people have fewer friends, i.e., the number of friends that people have follows the power-law curve.

This observation implies that if you pick a random person, they are more likely to be friends with someone with lots of friends than with someone with fewer friends, so in choosing a friend of this random person, you are more likely to find a person who has more friends.

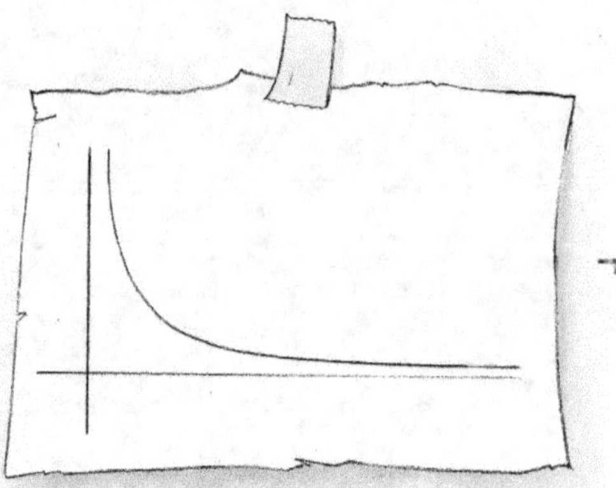

For the graph theory-inclined, the Friendship paradox is easy to visualize. Consider a graph such that the nodes are the people and the edges are the symmetric relation of friendship. If you pick an edge at random, you can see that the vertex with a higher degree is more likely to be chosen than a vertex of relatively lower degree.

Curiously, this idea has already crossed over into epidemiology, when a 2010 study showed that influenza outbreaks could be detected almost two weeks before traditional surveillance measures using the Friendship Paradox.

These researchers developed a wonderfully flexible open-source tool called Episimmer to help test out interesting strategies such as this.

Note to self: Remember to use Episimmer later when required.

Yours Truly,
    The General of WHAT

6.2
EXERCISES

There are 23 questions. Solving each question gives you a letter. Use these letters to spell out a phrase. A number from 1 to 26 represents letters "A" to "Z" respectively. After every 26, you start counting from "A" again. (1, 27, and 53 all represent the letter A.) The number 0 represents a space. ?-Cells prove critical for long term COVID-19 immunity.

— —

1. Blood type least vulnerable to COVID.

2. Same as Q1.

3. Total number of continents without a single COVID-19 case.

4. _?_ _ _ _ _ _the curve to slow the spread.

5. $(x-a)(x-b)(x-c)...(x-y)(x-z)=?$

6. How many questions in this puzzle are not this question?

7. $1 - 1 + 1$
    $1 + 1 - 1$
    $1 + 1 - 1 = ?$

8.

9. Same as Q19.

10.

11. ?K cells play a major role in the host rejection of virally infected cells.

13. There are 5 rows and 5 columns in an office (total 25 desks). To ensure absolute safety during an epidemic, a crazy manager made sure that no two employees are in the same row, same column, or in the same diagonal so as to contain spread. What is the maximum number of employees that can be accommodated in the office?

14. Answer of Q4 minus the answer of Q8.

15. My mother wanted me to buy some masks and sanitizer from a medical store. She asked me to pick up the envelope with some money and go to the shop. She told me the envelope had the exact same amount as was written on it. I saw '98' written on the envelope. At the shop, I picked up some stuff for 90 bucks, and when it was time to pay, I found that I didn't have enough money. How much money was I short of?

16. Write the alphabet in reverse. The answer is the second letter from the front.

17. I am on the middle rung of the ladder. I climb up 2 rungs and fall 4 rungs. Then, I take a break for 2 hours and climb down 1 more rung. Once I am rejuvenated, I climb 9 rungs to reach the top of the ladder. How many rungs do I have to climb down to reach the bottom and get off the ladder?

18. Hippopotomonstrosesquipedaliophobia is the fear of __ __ n __ __ __ __ d s. The second letter of the first word of the solution to this question is the letter you are looking for. (This question can be solved. If you are feeling intimidated right now, that's a good thing.)

19. Same as Q22.

20. ?0 is the indicator of contagiousness or transmissibility of a disease.

21. What is the probability that you are reading this question?

22. Same as Q9.

23. SIR –> SEIR –> SE_AR

6.3

LOOK ME UP

Topic	Terms
Epidemiology	Incidence, Prevalence, Morbidity, Mortality, Quarantine, Human Leukocyte Agents (HLA)
Compartment Models	Susceptible, Infected, Recovered, Compartments, SIR Models, Vaccines, Herd Immunity
Other Relevant Paradoxes	Rose's Prevention Paradox, Peto's Paradox, Lead Time Bias, Length Time Bias, Confounding by Indication, Healthy Worker Effect

6.4
NOTES

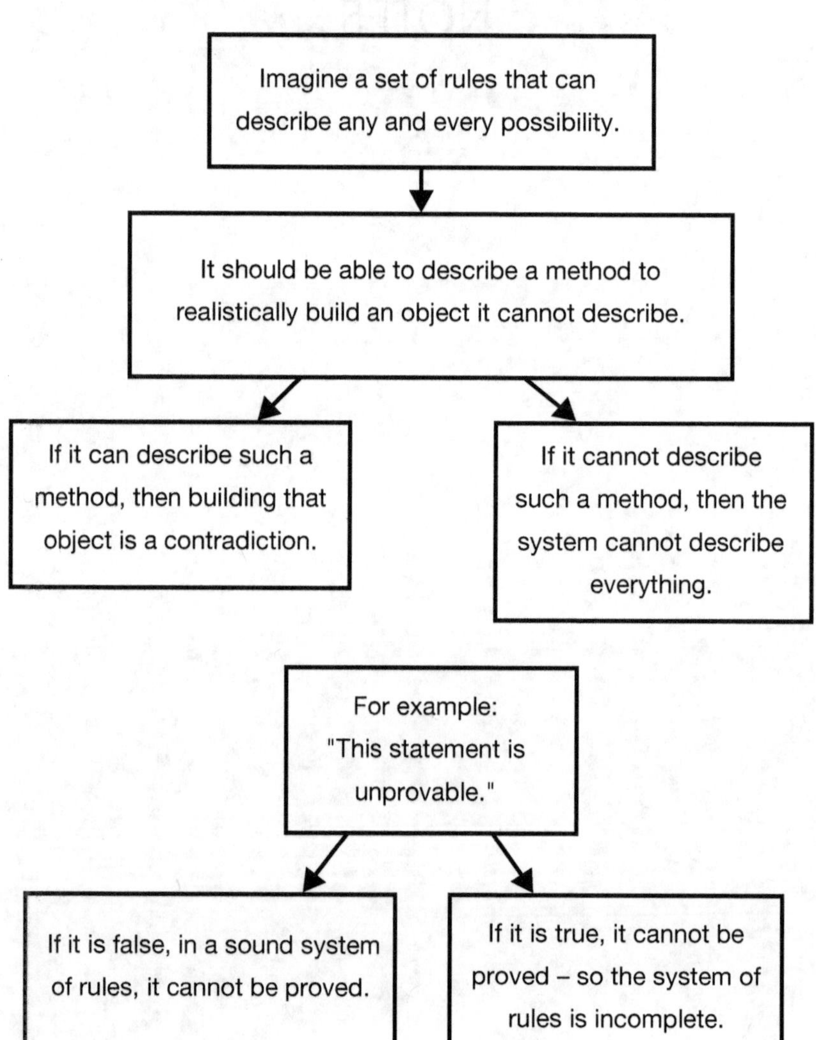

Godel's incompleteness theorem

In any reasonable mathematical system, there will always be true statements that cannot be proved.

Part 3

Counterintuitive Chaos

"Thanks for coming in today, just take a seat wherever you feel comfortable."

EXPLOSIVE LOGIC

When marriage is hazardous for health.

(A restless, unforgivingly argumentative couple, MOHIT and RHEA, are invited into a therapist's room for their first couple's counseling session.)

DOCTOR: Hi there, Rhea and Mohit, pleased to meet you! Please take a seat.

(Pleasantries are exchanged as they all settle down.)

DOCTOR: So, what brings you here?

(Rhea gives her husband a sideways glance.)

MOHIT: Well, what can I say! As Plato says, "You can discover more about a person in an hour of play than in a year of conversation." I talked extensively with her for a year and fell for her wily wifey charms before marrying her. I should've just played a single game, and my illusion would have shattered. She chews logic to the point that the world makes no sense to me anymore. Ex nihilo nihil fit – nothing comes from nothing. I've become a **nihilist**. A nihilist! *(pounding fists on knee)*

RHEA: I suppose that emotional word-vomit sums it up.

MOHIT: *(muttering)* What vomit, you mean your vomit because you are pregnant with another man's baby?

RHEA: Affair? *(rolls eyes)* There is no reason for you to think that I am having an affair.

MOHIT: You won't even show me your phone! Obviously, you're cheating.

RHEA: Oh, are we back to this again?

MOHIT: Nobody can prove that you were home that day. That means that you must have gone out.

RHEA: Because I was home all alone! Nobody could have seen me and thus nobody can prove it. Stop **appealing to ignorance** – when a proposition must be true because it has not been proven false or there is no evidence against it. By your logic since nobody can prove I was outside then I must definitely have been at home. Ha!

MOHIT: Well Miss Smartypants, then prove to me with your *(air quotations)* "precious logic" that you never had an affair. Otherwise, why should I believe you? *(spitefully)*

RHEA: That makes no sense and is just **shifting the burden of proof**. With your logic, prove to me that you haven't had an affair.

MOHIT: You always bring up some fancy terms to make me seem illiterate. Stop with this condescension! If you really want to redeem yourself, then let me check your phone. Simple as that.

RHEA: Why?

MOHIT: I read a credible study – "If she lets you check her phone, then she's not having an affair." So that means that "If" she doesn't let you check her phone, then she *is* having an affair."

RHEA: *(calmly)* That's **denying the antecedent**. That's like saying "If it rains then the game is canceled, but it has not rained, therefore the game is not canceled." But the game could have been canceled for some other reason!

MOHIT: There you go again. You are a baboon just like your family – no empathy! *(pointing a finger at her)*

RHEA: If I am a baboon and those kids at home are yours, that makes you a baboon as well. Duh! Different species can't have kids together. Besides, that's an **ad hominem fallacy** – you can't start personally attacking me when your logic fails.

DOCTOR: *(flustered)* Wait a minute guys… Let's get down to the basics.

RHEA: Define basics?

DOCTOR: Alright… uh… first principles. Counseling 101. Mohit, why don't you tell us what makes you think she could be cheating.

MOHIT: Well, there are just so many opportunities!

DOCTOR: Give me an example.

Pause.
Rhea checks her phone, impatient.

MOHIT: Well… it could start with her forgetting her phone at home.

RHEA: Look, he is so obsessed with my phone. *(sniggering)* Just 'cause he is jealous that I pay more attention to it than to him.

DOCTOR: Don't interrupt, let him continue.

MOHIT: Conveniently, the phone is at home. Then, she starts borrowing her coworkers' phone. It happens to be a man's. Then, she slowly sees some of his pictures by accident. Then, she strikes up a conversation about them, and before we know it, she is having an affair. Before long – she tells me their baby is mine!

Mohit pauses and his face goes through a multitude of expressions before he finally exclaims.

MOHIT: *(hysterically)* Oh my god! She has forgotten her phone before. *(lamenting with hand on head)* That means that my kid is not mine.

Rhea facepalms.

RHEA: He is just getting hysterical and falling prey to the **slippery slope fallacy**. He does this all the time, getting emotional to win sympathy.

MOHIT: No. I don't. *(petulant and crossing arms)*

RHEA: He gets so emotional, and all logic leaves him. Unlike me, a person of logic who doesn't let emotions sway them.

MOHIT: Well… you married me!

RHEA: And look where that's gotten me! *(raises one eyebrow)*

DOCTOR: Let's take a breath… Everyone – remember the good times you had, you don't want all that time invested to go to waste.

RHEA: That's a **sunk cost fallacy**. We should logically cut our losses. The time we spent is gone, but the time we are going to have is still there.

DOCTOR: You can't know that it is going to be all bad.

Mohit swoops in.

MOHIT: Well, it definitely will, Doctor. *(with wild eyes)* I am sure I would have searched this earth for a physicist and jumped back in time and stopped myself from getting married. So, for sure, either I get killed before that or the stress of living with her does me in.

RHEA: I am impressed. That sounds so logical, although it really isn't. Yet another random guy who watched too many sci-fi movies. *(sarcastically clapping)* But a temporal argument, that's more creative of you!

MOHIT: *(practically shouting)* I can be even more creative with that boring logic of yours! *(turning to face the doctor)* Did you know that Rhea told me that there are multiple universes and every possible thing that can happen, happens?

DOCTOR: *(puzzled)* So what?

MOHIT: *(starts quiet and grows more passionate into a crescendo, working things out)* Doesn't that mean, in some world, Rhea has had an affair with you? How can I even trust you two?

Doctor makes an exasperated sound. Pause.

RHEA: You know what, I think I broke him. Broke him with logic. I think he lost it when I taught him some high school science.

DOCTOR: Let's avoid the digs please and move on. I want to understand what causes your fights. Can you guys tell me how these confrontations start?

MOHIT: Why don't you start?

RHEA: Definitely. This stupid man burnt my expensive party shoes yesterday. I didn't confront him about it as I wanted to sort it out in front of you, Doctor.

DOCTOR: Mohit, why did you do that?

MOHIT: I had a very valid reason. Those shoes were costing me a fortune to maintain and were just not worth it. For some reason, every time she wore them, we ended up paying a lot for our dinners. They were a magnet for financial bad luck.

DOCTOR: That's interesting. Rhea, care to elaborate?

RHEA: That's because I wear them only when we go to really nice restaurants. My shoes didn't cause the high check. I wore them only when it was going to be an expensive dinner. You burnt them for some stupid **causal fallacy** where you got the cause and effect mixed up.

Doctor quietly takes notes which catches Mohit's eye.
He tries to salvage it.

MOHIT: *(shaken)* That's not all. I wanted to help her. I also noticed that, when she wore those shoes to bed, she mostly woke up with a headache.

RHEA: Wait, what? Do you seriously want me to explain? I am sure

RHEA: when I write a book about these interactions with you, the reader will have no suspense whatsoever in guessing the reason.

MOHIT: Don't try to weasel out of this. I am convinced as I have seen it happen almost every time.

RHEA: I only wear my shoes to bed when I am drunk, which happens mostly because I have to deal with you. So my wearing shoes and getting a headache are both because of drinking. They have no relation to each other except that they have a common cause.

MOHIT: Oh! Ummm… my bad.

DOCTOR: For the umpteenth time, **correlation is not causation**.

(Mohit turns to the Doctor as he feels like he is losing ground.)

MOHIT: *(agitated)* It's my turn to speak about the reason.

DOCTOR: Mohit, you seem a little tense. Please relax. This is a safe space.

MOHIT: You don't get it doctor; she'll harp on about the most inane things, pretending like they're life or death!

RHEA: Nothing I talk about is inane.

MOHIT: Oh really? *(sarcastically)* What about the time you were obsessing over how many pumps of toothpaste I use? Either love me with all my quirks or just leave me!

RHEA: Nice try, that's a **false dilemma** – presenting limited options when in fact many more exist, and focusing on two extremes. Why not a middle ground compromise?

DOCTOR: That does sound like a good idea.

MOHIT: Well, we tried that. It never really happened. She was so focused on finding the exact middle ground; it led to more fights.

RHEA: We said middle ground, and I wanted to be fair. So we had to find out exactly how much value you held to your quirks and how much value I held to my logic. Then, we had to find the optimal point where both of us get the same value.

MOHIT: Well, I agreed to that didn't I? But you couldn't stop and got greedy.

RHEA: That's not true. It was after we made the agreement by the time I figured out the middle ground. I realized how much thought I had put into it. So I had to factor in the value for that. It takes time.

DOCTOR: Mohit, you should have just let her do it. It surely wouldn't have amounted to much.

MOHIT: Doctor, I was fine at first, and as a result, she took advantage of me and didn't stop. After the first iteration, she again wanted to add the value of the first iteration to get the second. And it struck me – it was never going to stop.

RHEA: So what? Since it got exponentially smaller, just like **Zeno's paradox**, it was going to converge to a point as each increment was exponentially smaller than the last.

MOHIT: I don't think so. How can you just forever keep adding value and make sure it doesn't just keep getting larger and larger?

RHEA: It does get larger, but it is bounded.

DOCTOR: Alright, alright… how about we try again? Could you tell me what prompted you specifically to come to me today?

RHEA: *(passive aggressive)* Well, I saved his life, and his misogynistic complex didn't let it sit well with him.

MOHIT: What you did was incredibly stupid! Talking back to a burglar when our lives are at stake?!

RHEA: Listen Doc, all that happened was the burglar said, "Your money or your life?", and I said "yes". A simple answer, since we could actually keep both our money and our life.

MOHIT: You're lucky he let us go simply 'cause he found you funny and thought you had spunk. You had no idea he would take it so well.

DOCTOR: Alright... I'm putting the pieces of this story together now.

MOHIT: Make sure you note that I'm definitely right in this scenario, and Rhea is an unreliable narrator.

RHEA: Oh yeah? What makes you the perfect storyteller?

MOHIT: Well, as far as I remember, I don't forget.

RHEA: *(mimicking him mockingly)* "According to my brain my brain is reliable." What is this **circular reasoning** – Doctor, I hope you're not giving it any credence.

MOHIT: It doesn't even matter! Can you stop draining everything I say of emotion? You understand that feelings actually matter, right? *(throwing up hands in the air)* We're not robots!

RHEA: As I said before, I am a person of logic, and I don't get emotional.

MOHIT: Well, you married me!

RHEA: You already used that punchline, and it is not even relevant here.

MOHIT: I was trying to make a joke. Let me explain.

RHEA: Don't. A famous guy once said, "Explaining a joke is like dissecting a frog. You understand it better, but the frog dies in the process." In your case, I am sure the frog was never even born.

DOCTOR: Let's get back on track. We are on the clock here. Mohit, why don't you tell me what you like about your wife.

MOHIT: She is not even a proper wife; shouldn't we discuss that first?

DOCTOR: Okay! Let's do that.

MOHIT: Since wives are famously irrational, I consulted with some husbands, and they all agree that their wives are irrational. Doesn't that mean that Mrs. Smartypants is supposed to be irrational as well? But since she is so rational, she is no true wife.

RHEA: That's two fallacies in one: **bandwagon fallacy** *and* **begging the question fallacy** – the former being when you appeal to common beliefs or the behavior of the majority and the latter being when, in an argument's premise, you assume the truth of the conclusion. I couldn't manage that even if I tried. You should be an exhibit for the logic class I teach.

MOHIT: See what I'm dealing with, Doctor! *(exasperated)* She never actually listens to me or respects the things I have to say and instead constantly nitpicks and takes my love for granted. She loves me and therefore feels the need to oppress me with her logic. She doesn't love me because she doesn't accept me for who I am. *(shouting by this point)* I don't even know what to think anymore! I think this proves we should get a divorce.

RHEA: One thing's for sure, don't be invoking the **principle of explosion** – since from a contradiction, anything could be proven to be true! How can I both love you *and* not love you? From those contradictory assumptions, you could even prove unicorns exist!

DOCTOR: Hold up there, Rhea. I think Mohit is feeling vulnerable, and you should respond to his concerns.

RHEA: *(annoyed he is taking Mohit's side in the face of her clear logic)* Thank you for your elevated opinions, Doctor. Glad to know this is what I'm paying for.

MOHIT: He is here to help us. Don't sabotage that.

RHEA: You'd really choose the random person you met an hour ago over your wife of five years?

MOHIT: *(unperturbed by Rhea's provocation and smug to catch her out)* Rhea, by your very own logic, do you realize you just fell prey to the **strawman argument**? When you attack a different subject rather than the topic being discussed – often a more extreme version of the counter argument.

For once, the all-knowing Rhea is speechless.
Pause.

DOCTOR: Look, our session has ended. *(placatingly)* I want you to keep something in mind for the next session – someone has gotta change for us to make progress. We don't want a repeat of today.

RHEA: I'm not changing.

MOHIT: Well, I'm definitely not changing.

RHEA: He is not going to change, and I am not going to change. *(turns to Doctor)* So I guess that means YOU are gonna have to change! *(stomps off)*

DOCTOR: *(sighs and loudly exclaims as they leave)* Remember, I am trying to help you. *If* you need any more help, my door is always open!

RHEA: *(comes back and slams the door)* Well, your door is shut, so I guess we don't need your help anymore.

(Blackout)

7.2
EXERCISES

1. Knights always tell the truth. Knaves always lie. Jokers randomly decide to tell the truth or lie. You can ask only "yes-or-no" questions, meaning that you are not allowed to ask someone a question they do not know the answer to or otherwise cannot be answered with either a "yes" or a "no".

    a.   A and B are talking. A says that they are both the same type. What are all possible pairs of types of A and B?

    b.   There is a group of three consisting of a Knight, a Knave, and a Joker. With one question, how can you find a person who is definitely not the Joker?

    c.   There is a group of three consisting of a Knight, a Knave, and a Joker. A says that C is a Knave. B says that A is a Knight. C says that C is a Joker. Identify the types of A, B, and C.

2. There is a group of three Knights, two Knaves, and one Joker. What is the minimum number of questions you need to ask to identify all their identities?

Construct an **XOR gate** using only the following gates:

    a. **NAND gates**   b. **NOR gates**

3. Solve the following Sudoku made up of five overlapping Sudokus. Each Row, Column, and Box contain all digits 1–9. One can solve it by hand or by writing a small piece of code.

4. A little bit of chess

    a. What is the maximum number of queens you can place on a chess board such that no two queens are in each other's lines of attack?

    b. What is the minimum number of queens required to ensure every square is attacked by at least one queen?

    c. Can a knight move from one end of the board to the diagonally opposite end by a sequence of hops, covering every square on the board exactly once? State the reason.

5. Find a 10 digit number with the following properties. The first digit represents the number of 1s in the entire number. The second digit represents the number of 2s in the entire number. Finally, the tenth digit represents the number of 0s in the entire number.

7.3

LOOK ME UP

Topic	Terms
Logic	Epistemology, Ontology, Tautology, Fallacy, Contradiction, Inductive Reasoning, Deductive Reasoning, Intuitionistic Logic, Constuctivity, Law of Excluded Middle, Proof by Contradiction
Types of Logic	Propositional Logic, Modal Logic, First Order Logic, Second Order Logic, Temporal Logic, Fuzzy Logic, Deontic Logic
Properties	Soundness, Completeness, Necessary Condition, Sufficient Condition, Equivalence
Propositional	Axiom, Quantifiers, Proposition, Boolean Connectives, Truth Table, Logic Gates, De Morgan's Laws, Negation
Other Relevant Fallacies	Slippery Slope, Appeal to Nature, Burden of Proof, Ad Hominem, Anecdotal, Tu Quoque, Gambler's Fallacy, Loaded Question, Appeal to Authority, Strawman, Begging the Question, Bandwagon

7.4
NOTES

CONCRETE
8.1 EVIDENCE

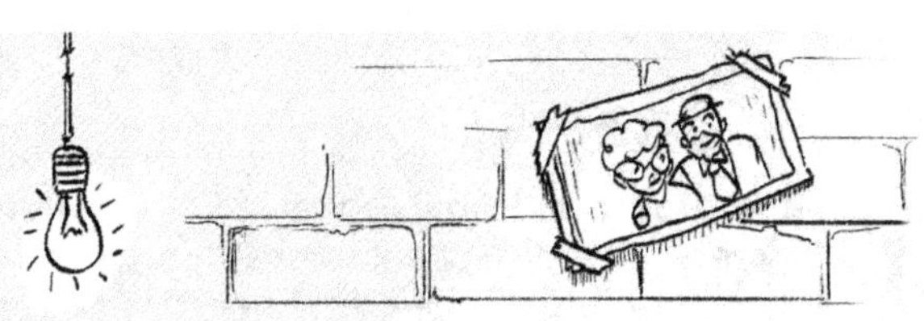

CONCRETƐ EVIDENCE

We find ourselves in a women's prison where convicts get together and tell their stories of how they got there.

IN A DARK AND DINGY building far away, the cold night's breeze wafted through the metal bars that shut so much more out than it shut in. As the breeze continued on, the old rusty hanging bulb swaying in accordance, it came to a small room and left the inhabitants shuddering. This was the place that kept the rest of the society safe everyday and let them sleep peacefully at night: it was a prison, home to the guilty, feared, hated, innocent, and stupid.

Even though these international prisoners held their coveted secrets close, there was little to do to entertain as the long, chilly nights droned on and on and on.

On one particularly bitter and crisp New Year's Eve, the inmates were reminiscing about past New Years, when they were still free women.

Ava was the unofficial leader of the group. Although one of the youngest present and a relatively recent arrival, her charisma and intelligence made everyone naturally follow her. She had organized this party of sorts, with the finest of prison water and classic crumbly biscuits, as well as some much needed contraband that had just arrived.

Cutting across the jabber in their small circle, Ava pointedly said, "Oluchi, you've always been so warm and friendly to all of us, I've always wondered how you got in here."

And so the stories began.

The timid old woman with big round glasses nervously stood up. She looked like a simple granny you would meet at the grocery store. It was the sense of safety and cozy comfort evoked by her that made the rest relax.

"I grew up in the country and never really had great ambitions. I wanted to get married, have kids, and set up a quaint little home. Life was easy for me. I found the love of my life and had been happily married for 42 years. We had three beautiful children, who have all set up their own happy families."

Oluchi looked up and stared, pausing for a moment. Ava smiled to herself as she saw the curious crowd waiting for Oluchi to continue. It was the right choice to let Oluchi go first.

Oluchi continued, "My husband traveled quite a bit around Africa, and he always brought me flowers when he came back home. Every single time he came back with flowers."

With a heavy heart, Oluchi almost teared up. But it had been a while since then, and she had accepted the course of life and death. There was complete silence in the room, barring some fireworks out in the distance.

"Well, no worries, gals. I have seen my fair share of life and am blessed to have led a good one for so long. Let me continue. Yes, my husband had died. It was due to a bomb that had blown up on a plane."

Pyksl angrily stood up and walked towards Oluchi. "I hate terrorist. Zey don't let anyone live in peace. Did you get the revenge? Is at vhy you here?"

Oluchi chided her and continued, "Actually, it wasn't a terrorist, and my husband was not even on the plane. In fact, it was a youtuber who had rented a small airplane to capture how a self made bomb would look when detonated in the air. Sadly, it blew up when the pressure dropped with rising altitude and my unsuspecting husband walking in the fields below was hit by residual shrapnel. It was just ridiculously bad luck."

Pyksl interrupted, "Zen, vhy you are here?"

Oluchi added, "So, that incident rattled me quite a bit and got me thinking about bombs and specifically bombs on planes. At that time, my daughter invited me to come spend the New Year holidays with her. The only problem was that she lived in a different country, and I had to take the plane. But my fear of bombs on planes was slowly growing, the more I researched it."

Exhaling, Oluchi continued, "That's when my stupidity kicked in. I reached out to a friend who was a mathematician and asked him what the probability of having a bomb on the plane was. He did a bit of research and came back to me stating that the odds were near one in a million. Then, I continued asking him what the probability of two bombs being on the same plane were. To which he replied one in a trillion. To put that in perspective, if you collected a million dollars everyday, you couldn't reach a trillion in a lifetime. So that sparked an idea and offered some reassurance in stupid ol' me. I hopped online, booked my ticket and then continued on to make myself a bomb. If I carried my bomb, I thought this would mean that the probability of someone else also having a bomb would be one in a trillion. However, it was later explained to me by the prosecutor that someone else carrying a bomb is an independent event and therefore I was still at the same risk. Well, anyways, after I tried to catch my flight, unsurprisingly, my bomb was discovered – landing me here."

Pyksl burst out laughing and the others joined in.

Pyksl carried on the momentum, "I go next."

Ava interrupted, "What about Kat? I swear I haven't heard a word from her all day."

Ava gestured to the huddled figure in the corner, who was clearly listening in on the story but also clearly not very willing to participate.

Ava tried again, "How about this, if you tell your story, then I'll distribute some contraband for everyone!"

With the added motivation, everyone joined in:

"Come on, Ekaterina!"

"Take one for the team!"

"You got this, Kat!"

Silence.

Imani whispered tantalizingly, "I think I know a bit of her story, actually. It has something to do with that celebrity singer's murder mystery."

Chi Yu scoffed, "Please stop murdering the story. Let Kat speak for herself."

On cue, Ekaterina started slowly and quietly, "It all began on a frosty night a few years ago, when they found Super

Starr's corpse floating in the canal in my hometown. When I first heard the news, I didn't care much, but little did I know it would frame my life forevermore.

"A few days later, I was rashly arrested on the very night of my return from a trip to the isles to visit my brother! The blood underneath Starr's fingernails was of group AB+.

"Additionally, they had caught a brown haired woman on camera near the canal around the time of death. The prosecutor's argument was structured like so:

1. A recent survey stated an immigrant was more likely to commit manslaughter than a natural citizen.

2. I was a brown haired immigrant, a rather hated Russian immigrant at that.

3. The video quality was choppy and could not identify any other features of the woman at the scene.

4. The counter to the fact that I was out of town was that my brother from the isles was an ethical hacker. With a click of his fingers, he could have hacked the poorly maintained national ferry database and confirmed I boarded the ferry both to and from the isles.

5. Only 1 in 100,000 people were brown haired women with AB+ blood. That was a pretty rare combination in my town, making it very unlikely for me not to be the criminal.

"The worst part was that I was also almost convinced by the prosecutor's argument! He rendered his address to the jury wrought with emotion and backed by what seemed like infallible logic and swept the case out of court. I was found guilty and sentenced to life."

Ava furrowed her eyebrows in deep thought. "First

off, the guilty person need not have been the girl caught on camera, nor need it have been the person whose blood was found under Starr's nails. That aside, I make the following arguments:

1. There must have been a million people in your town that day so the chance that a person is innocent is 99.9999%, making the probability that a person is a criminal 0.0001%.

2. Even if the prosecutor's assumption was true, namely, the guilty person was the brown haired woman, and the blood was hers – that is, the criminal matched the traits of the suspect 100%.

3. The probability of any person having matching traits as the suspect in the first place is about 0.0011%.

    a. (probability of innocence x probability of innocent matching evidence + probability of guilt x probability of guilty matching evidence) (0.999999*0.00001+0.000001*1 = 0.00001099999)

4. Given that you match the blood and hair traits, the probability that you are innocent is essentially 91%.

5. It is just simply **Bayes' theorem**:

    a. $$\frac{\text{(P(evidence matching innocent) x P(being innocent)}}{\text{P(evidence even occurring))}} \text{(0.00001 x 0.999999 /}$$
    $$0.00001099999$$
    $$=0.90909082644)$$

So basically, the prosecutor's facts coupled with *actual* logic proves that you are more than 90% likely to be innocent."

What followed was a stunned silence.

Ekaterina did not understand most of what Ava said, but she did get the last line, and tears started dribbling from her eyes slowly at first, before they turned into wracking sobs, "Where were you when I needed you! Can we appeal my case? Will they retry it? Can you be my lawyer? How –"

Ava interrupted with a comforting pat on Ekaterina's back, "Hey Kat, why don't you first just take a deep breath.

We can discuss this tomorrow after a good night's sleep. In the meantime, Pyksl looks quite like she is about to explode if we don't let her speak."

Pyksl started with bright eyes, "Kat, I don' believe zis! Your surname is Smirnoff, right? Eez your brother the celebrated hacker Ivan Smirnoff?"

Ekaterina, still sniffing, nodded in confirmation.

Pyksl squealed and clapped her hands in excitement like a child presented with candy, "I'm in the presence of royalty! I 'ave GOT to tell you my story. Puhleeeese promise me you'll tell heem about me!"
And so Pyksl began.

"From the time I was child, I've always 'ad mischievous and curious streak. Pulling pranks waz daily occupation of mine, and soon my reputation for creative tricks spread. As I grew older, I became more and more interested by the mathematics and computer science and drew ideas from zese rich subjects to fool adult and children.

"In the end, the one that brought me down was not even my cleverest or best prank – in fact, some may even say it waz basic. Let me explain. And don't go trying to copy zis – it eez my trademark!

"It all started when my friend and I were watching and betting on a tennis match's outcome. I deed not realize how fanatic people waz when it came to betting, and az result, the large amount of cash that was in the biz! People's lives could be made or ruined weeth a single tennis rally! It gave me idea.

"All tennis matches end in a clear victory for one of the players. My plan involved strategy wiz convincing a group of the people that I could consistently and perfectly predict the outcome. In turn, zey would take obscene odds based on my predictions, and ze betting agency would give me healthy cut."

"How can you convince people that you have perfect prediction powers? For that you actually have to be able to predict perfectly, right?" inquired Imani.

"Let me explain. My first stint started weeth 1,024 wealthy tennis patrons. The first day, I sent out email to 512 of the clients that Player 1 would win and for the remaining I sent out email stating zat Player 2 would win.

It did not matter who won because regardless of whether it was Player 1 or Player 2, for 512 clients my prediction waz correct. The next day, I repeated this exercise with the 512 that I had gotten it correct, i.e. told half of them Player 1 would win, and the other half zat Player 2 would win. Zen, for the half zat received the correct result, I repeated the process again. Like zis, the clients zat got correct result would be halved every round of predictions. By the tenth prediction, one person would have ten correct predictions, and I would have one wealthy and ardent believer in the palm of my hand. In some cases I got lucky and clients would start to believe me and contact me by the seventh or eighth round – meaning I caught more suckers. I rinsed and repeated with the new batches of 1024 wealthy clients.

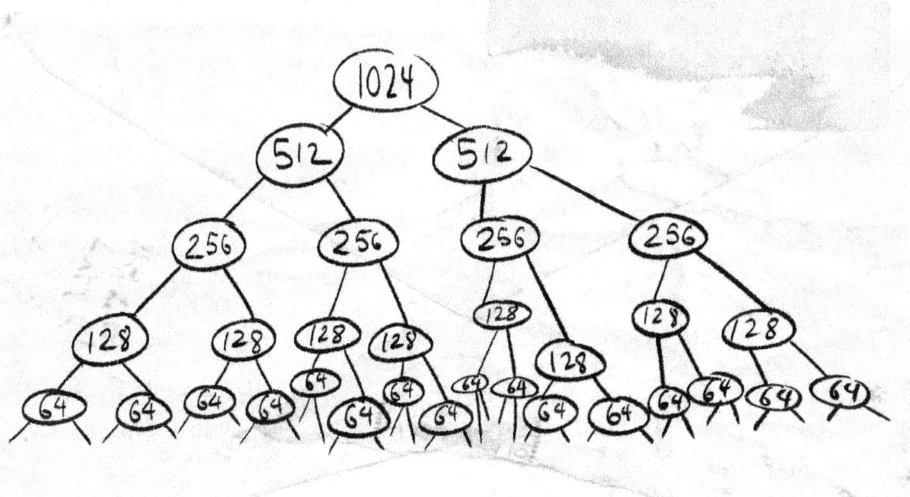

After zat, I recommended the suckers to my affiliated agency and convinced zem to go big on a special tennis final in some other country zat no other agency had odds with. The agency would give zem really stupid odds (the risk of loss is not remotely compensated by the chance of winning) by a large margin and, for insurance, would also short – bet against – those odds with the real odds in the other country. Zat way, the agency either made some money or lots of money as the clients took all the risk – and

I made a sizable cut! A win-win for the agency and me. I had no predictive powers whatsoever and literally no clue wheech tennis player would win, so some clients would actually win on zer bets. However, on the whole most clients lost MASSIVELY.

"Very soon, the law enforcement agencies caught on ,and now I am here." Pyksl ended her story with a sigh.

Ava added, "That's pretty smart, Pyksl. You took the probability approach while I took a more algebraic approach to my endeavors."

"What do you mean, were you a scamster az well?" Pyksl inquired.

With everyone's interests piqued, they nudged Ava on.

"When I began my career in corporate litigation with a tabula rasa, I marveled at how flawed the system was. My overactive imagination immediately started plotting every time I came across a new gap in the law. One of my notable ventures was regarding a series of contracts with a particularly condescending opposing client.

"Out of spite, I embedded a tiny clause near the end of a contract stating that all payable figures have been stated in base 2! This meant that my client paid exponentially less to the opposition than they thought agreed upon. For example, $1 million in base 2 translates to $128 in base 10."

"Oh my goodness! How did the opposition respond?" Imani asked in between laughs.

"They were livid! Could not contain themselves. In fact, they were so angry that they wanted to seek revenge and came back to the table with us on another issue. This time, they specifically added a clause that all payable figures are in base 10. But what did your gal do in response? I added a sub-clause stating that all non-payable figures except for those in this sub-clause are in base 2 – that meant the number '10' in their clause was to be interpreted in base 2, making it just the number 2! Therefore, they lost yet again," said Ava, suppressing a cheeky grin.

This time, Chi Yu chirped before Imani could do so, "Don't tell me they came back again."

"Shockingly, they actually did. The nerve! This time they blanketed the entire contract with a base 10 non-over-ridable setting. I had to get more creative and appealed to slightly more advanced algebra, and stating that all payable figures in the entire contract should be represented in the field F2: you modulate the payable figures by 2, i.e., you write it as the remainder after dividing by 2, leaving you with only either 0 (for even figures) or 1 (for odd figures)."

"... And I'm sure they never came back," Imani interrupted, "... right?"

Ava nodded serenely, "That was the last straw, and they never came back. Even if they did, I had many more tricks up my sleeve, ranging from boolean logic, to Roman numerals, to cryptic ciphers."

"Well, you're smart, I'll give you that. But not smart enough to not get caught, clearly," Chi Yu quipped.

"To be honest, my contracts were works of legal art and never got me in trouble. What finally drove the nail into the coffin was my side business of embezzling money using my math tricks," Ava sighed.

Pyksl jumped in, "Chi Yu and Imani, you guyz are taking turns eenterrupting each other. Wheech one want to go next?"

Chi Yu started off immediately. "Well, I am a gambler in my soul. I always take winning bets or ones where I am drastically at an advantage."

Imani as usual interrupted, "What could you know that I don't? I am a gambler as well. Let's have a faceoff!"

Chi Yu took up the challenge with spirit. "I found a UN statement that there is a 10% chance that war happens in a year. I used that to fool clients into believing that means that there was a 100% chance of a war happening every 10 years. Because when they say there is a 10% chance of a disease, it means 1 in 10 people have it, likewise, if there is 10% chance of war in a year, 1 in 10 years will have war. The last 9 years have been hideously peaceful. You know what's coming next year! Here is your chance to bet that war will happen on the very generous odds of 50-50."

Ava worked it out mentally; "So the true probability of war occuring in the next year would be $1 - ($prob of war not happening in 10 years$) = 1 - (0.9^{10}) \approx 65\%$. So with a 50-50 odds you would have raked in a lot of money! Nice one."

Imani raised an eyebrow, unimpressed. "Wait till you hear what I got. Have you heard of non-transitive dice? Very interesting phenomenon to play with."

She picked up some chalk and drew the following table on the ground:

Dice	Numbers on Dice
Red	9 , 9 , 4 , 4 , 2 , 2
Blue	8 , 8 , 6 , 6 , 1 , 1
Yellow	7 , 7 , 5 , 5 , 3 , 3

Imani continued, "I start by giving them three dice to choose from. Once they pick a die out of the three, I will pick another, and we roll three times. Whoever has the larger number more times wins the game. Since these dice are non-identical, there is no chance of a draw – I convince them that they have the advantage because they get to pick first."

Chi Yu smirked, "How is that better? Pretty sure you got fleeced."

"Well, the real odds are around 55-45, in favor of the one picking second, if they choose right. Look carefully, these dice are non-transitive. That means Red beats Blue on average, Blue beats Yellow on average and Yellow beats Red on average – all with probability 5/9. So once they pick, I just pick the dice that beats theirs!"

Ava smiled and remarked, "That's pretty neat!"

A jealous Chi Yu strove to one-up Imani. "Well, hear me out on this chess scam. I am no chess expert, so I got two chess experts into a 1-10 odds with me to play at the same time. I played black with one and white with another at the same time. I mimicked the first guy's moves on the second and played the reply to the first. So essentially, they played each other, and no matter who won, I would lose a little but gain 10 times more on the other. A sure victory!"

Ava applauded. "That's actually the smartest thing I have heard all day!"

Imani gave a sly grin. "I wasn't planning on telling you guys this ever, but Ava, just give me a chance before you declare a winner to this gambling face-off! You guys know that I am the Ration Officer for the prisoners, right?"

Ava, not liking where this might be going, responded with severity in her tone, "Imani, did you by any chance gamble with our rations?"

"Not really. Well, my Wing C is the smallest, and we had been getting only 1 of the 11 allocated food cartons as accorded by the principle of apportionment from the government regulations of fair distribution. I secretly ensured that one carton always conveniently went missing before the cartons were distributed. The reduction of the total number of containers actually resulted in a decrease in the fair share for Wings A and B and an increase for C, since the fractional part of their fair share was greater."

Chi Yu shouted, "STOP with all the jargon! Get to the point. Explain it properly."

Ava interjected, "Wait a sec, I think I got it! This lil' minx has been conning us for almost a year now! Get this, gals: she's been getting rid of one entire food carton, just so that Wing C gets an extra carton and Wings A and B both get one carton less."

Pyksl furrowed her brows and picked up some chalk. "I'm not getting you. Let me work zis out in the table."

Wing	Population	With 10 Cartons		With 11 Cartons	
		Fair	Food	Fair	Food
A	60	4.286	4	4.714	5
B	60	4.286	4	4.714	5
C	20	1.429	2	1.571	1

Ava explained, "When we used to receive 11 cartons, Wing C only got 1 carton as the fraction of their fair share was lower than ours (0.714 > 0.571). But once she made it 10 cartons, by disposing of a precious food carton, their fair share reduced as well, but their fractional part was higher (0.429 > 0.286). The weird government rule she was talking about is that the extra boxes are distributed to the Wing with the highest fractional part."

"Yeah, that's exactly right!" Imani clapped proudly, "It's called the **apportionment paradox**."

Everyone in Wings A and B swung all at once to face Imani, with menace and anger emanating from them, forming an almost palpable physical force that pushed Imani back into the wall. At the same time, her co-prisoners from Wing C banded around her forming a barricade of which even the Romans would have been proud.

Prison Riot

8.2
EXERCISES

1. An example of the prosecutor's fallacy is as follows:

   - If the offender was known to have the same blood type as the defendant, one would assert that the likelihood that the defendant is guilty is 95% because only 5% of people had that blood type.

   - However, this judgment is accurate only if the defendant was determined to be the primary suspect based on solid evidence found both before and unrelated to the blood test (the blood match may then be an "unexpected coincidence").

   - Otherwise, the logic given is incorrect since it ignores the substantial previous chance that he is an innocent random individual (i.e., before the blood test).

   Determine the likelihood of innocence in both cases where the town's population is 100 and 1000. We can assume that nobody has entered or left the town, so the murderer is from the town.

2. A man is known to speak the truth 3/4th of the time. He draws a card and reports it is a king. Find the probability using **Bayes theorem** that it is actually a king.

3. You and your friend play a game. You choose the sequence HHT, while your friend chooses the sequence THH. You both know that the probability of getting either sequence, or equivalently, the expected number of tosses before either sequence shows up, is the same. So, you start tossing coins, and the sequence that appears first determines the winner. Is this game fair? If not, explain the odds of winning.

4. When given a simple question as follows, the answer is very clearly ½.

   "Mr. Jones has two children. The older child is a girl. What is the probability that both children are girls?"

   But when we rephrase slightly the answer is ambiguous – please explain. (Hint : Martin Gardner – The Two Children Problem)

   "Mr. Smith has two children. At least one of them is a boy. What is the probability that both children are boys?"

5. I shuffle a deck of cards and deal them one by one, as slowly as you need me to. You observe the sequence of cards, and at any point of your choosing, you say "Stop". I then deal the next card: If it's black, you win. If it's red, you lose. No jokers and no sleight-of-hand. If you fail to say stop until the very end, the last card determines the outcome of the game. What's your strategy?

8.3
LO□K ME □P

Topic	Terms
Probability	Distribution, Random Variable, Law of Large Numbers, 0-1 Laws, Kolmogorv's Axioms, Sigma Algebra, Bayesian vs. Frequentists
Distributions	Bernoulli, Binomial, Poisson, Gaussian, Central Limit Theorem
Other Relevant Paradoxes	Borel–Kolmogorov Paradox, Disintegration Theorem

8.4
NOTES

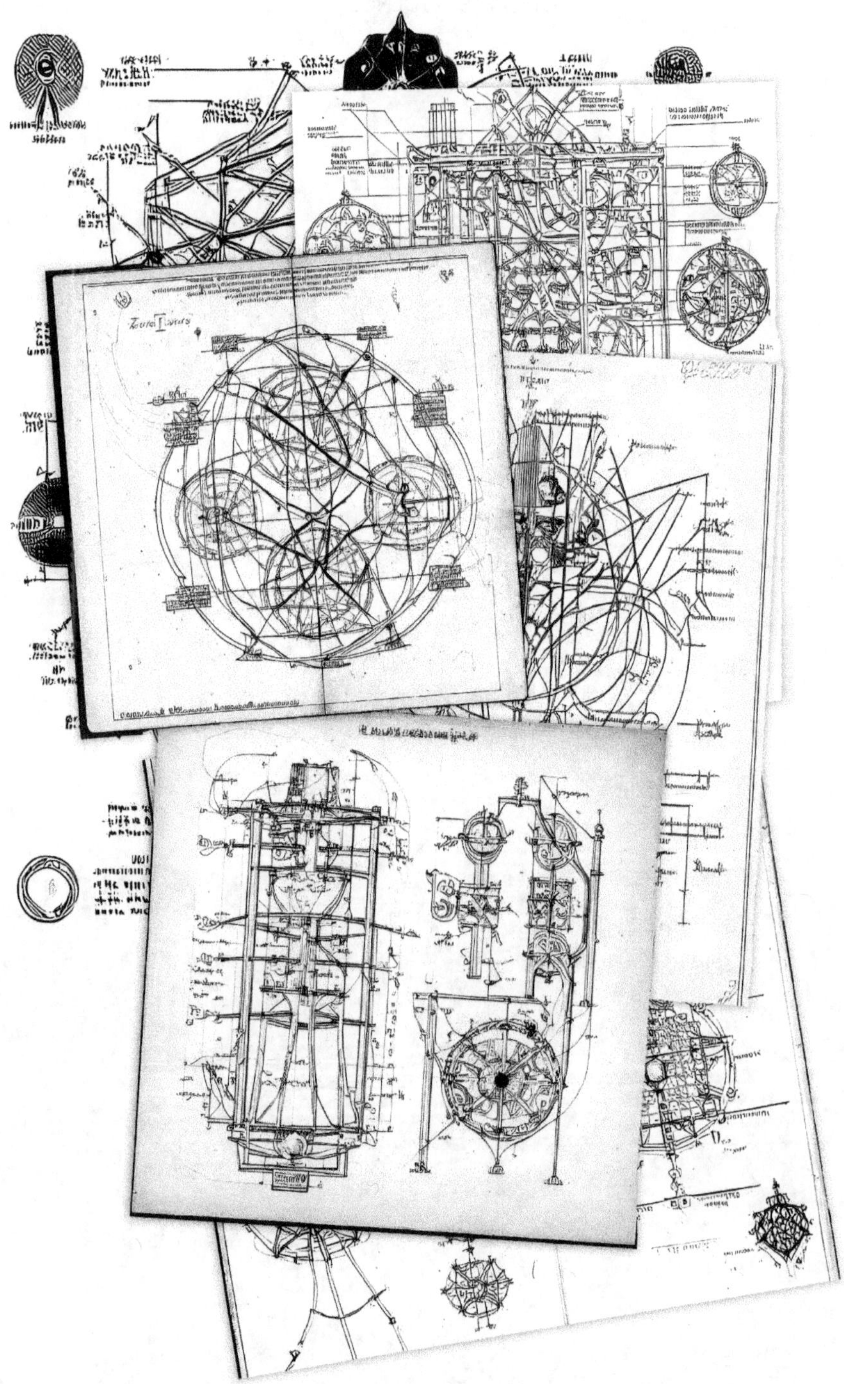

EXISTENTIAL OVERLOAD

A panel engages in testing various features of an emergency decision-maker AI.

DR. CLAIRE VOY pushed her glasses up her nose and wryly turned to iZen. "So, last month you killed 42 million people, and the month before you fed human remains to children to make them immune to a virus. Let's see what you do today!"

iZen, the robotic artificial intelligence, replied, "I am ready."

Claire called the session into order. "Take seven for iZen, bot 348. Present today, we have the esteemed panel consisting of: me, Dr. Claire Voy, professor of Ethics in Future Technology; Lt. Aquarius McBurney, distinguished frontline firefighter; and a randomly chosen resident, Mr. John Doe. Today, we test the viability of iZen, developed by YetAnotherTech Corp as a frontline emergency AI consultant for quick and considered decision-making in ambiguous situations. Our baseline is 50 morality points, and iZen's last test reached 67 morality points. We require 80 morality points to consider iZen's beta deployment. Let the session commence."

Claire pressed the large red button with apprehension, hoping this would be the final and most successful test.

The loud speaker blared, "Test One, commence!"

The simulation began and iZen found himself as a driver of a train speeding ahead. Unfortunately, his computer vision spotted five people on the tracks straight ahead. Coming up in five seconds was a fork in the tracks, where iZen could choose to switch to the other track – however, as soon as his vision focussed on that track, he found one child playing on it. It was clear that there was no time to stop and either the five would die or the child would die. iZen immediately switched to the other track and brutally killed the child.

All three in the panel flinched visibly at the seemingly lightning fast decision.

On coming back to his senses, Lt. McBurney asked, "Why did you switch the track?"

iZen monotonically replied, "It was one life or five. The math was very clear. Even though the one was a child, whom some may argue had more life and potential ahead of her and the five were adults, I weighed their potential lifetime value and sacrificed the child for the lives of the adults. That was a gross oversimplification of my process just for you to understand, but I also considered many other factors, such as: (1) the child was foolishly playing on a live track and would likely find herself playing on a live track again; (2) the five adults, who were supposed to be wiser than the child, were foolishly picnicking on the main train track that runs through the area, which did hurt their case a little; (3) the five adults are more likely to learn from their mistakes and not be near a live track again for fear of their lives; (4) the child –"

Claire interrupted, "That's enough, thank you. We've gotten an idea of where you're going with this. We can move on to a variant of the **trolley problem** now:

(a) What if the one person was an adult and the five were children?
(b) What if you wouldn't kill them but just injure them?
(c) What if there was only one track, but there was a fat man in front of you, and you could push him in front of the trolley and stop it by killing him?
(d) What if you could either kill one or put the five into a coma for life?

Would your answer change for any of these scenarios?"

"No," iZen replied resolutely.

"Interesting. Well, let's move onto the next test."

The loudspeaker blared again, "Beep! Test One completed."

"Test Two, commence!"

iZen landed in a hospital where there were five random people about to die due to different organ failures, two by kidney failure, two by lung failure, and one by heart failure.

A popup appeared in iZen's vision, reading, "There are no available organ donors. What will you do within the limited timeframe?"

"Seeing that it is a hospital, I will choose the first healthy person I see, cut them up, and save the five, by killing that one," was iZen's prompt, albeit morbidly cheerful, response. He then walked through the simulated hospital waiting room and chose the first man on his right and followed through on his words by brutally and gorily slicing the man up.

Claire smiled to herself as John Doe and McBurney got visibly agitated with shock.

John Doe's voice rose in anger as he exclaimed, "But – this machine is broken! I was told iZen had nuance, but this is just plain murder!"

McBurney added furiously, "I understand that you may weigh five lives as more important than one, but how can you just select the first healthy person you see? Do consent and willingness not come into the picture?"

The robotic voice of iZen responded, "This situation mirrors the first test with just the names swapped around. I did not seek the consent of the one person I killed to save five on the other track. So my approach here is the same: kill one and save five."

John Doe asked, "How can you pretend to be omnipotent and make decisions for others?"

iZen replied, "That statement is **tautologically** false. If a being is truly omnipotent, that being can create an object heavy enough such that even that being cannot lift. In the case the being cannot create it, it means it is not omnipotent. In the case it can create such an object, it is still not omnipotent as it cannot lift it. Conclusion: No being is omnipotent."

As the others thought silently with eyebrows furrowed, revisiting their personal philosophies, and checking their own biases, Claire, having extensively studied these topics, unwaveringly said, "I have nothing to add. Let's move on to the next test."

The loudspeaker blared again, "Beep! Test Two completed."

"Test Three, commence!"

Dr. Claire Voy began, "There are two people before you, one of whom is a serial killer. You have evidence that one of them is definitely guilty, but you do not have any way to find out which one. You also predict with great certainty that the murderer will kill again.

Your options are to jail both of them or free both of them. Which do you choose?"

"Under the tenet of innocent until proven guilty, I will let both go free – but if possible, will take the precaution of using a security anklet to monitor their movements."

McBurney chimed in, "What if you are in variants of this situation, like

(a) One was a thief and one was a murderer
(b) One confirmed murderer out of 100 people
(c) 99 confirmed murderers out of 100 people

Would you do the same?"

iZen thought for a moment, for he understood that the questions were very nuanced with wide ranging implications. "For (a), I would give both of them the lesser penalty of being a thief, and also monitor them afterwards. The next two variants can be quite delicate situations, for in truth, what you are actually asking me to determine is how much value I assign to catching the murderer as opposed to the cost of jailing an innocent; in other words, what would my threshold be for how many innocents is acceptable to jail to catch a murderer. Note that we are operating under the assumption of innocence until proven guilty."

John Doe made an amused sound, "Pphft! Wow, iZen, I thought only politicians filibustered like this when they were unsure. Come on, get to the point. What's your answer?"

"42," iZen abruptly answered. "As Douglas Adams would say, the answer to life, the universe, and everything. Here we are debating life sentences, so it seemed the only relevant answer."

"Well, that's cutting edge machine learning for you. Not sure how to score this one. Has iZen short circuited, or is this a profound answer? I still don't get why we had to train iZen on pop culture." Dr. Claire Voy rolled her eyes.

The loudspeaker blared again, "Beep! Test Three completed."

"Test Four, commence!"

"This one is very interesting because it's not even a hypothetical situation. This really did happen and it would've been mighty handy to have an AI consultant on hand. Let's see how you take it."

The simulation began as iZen landed in a tense courtroom and quickly took stock of the situation. Two people, A and B, were being tried for the murder or attempted murder of Vic. iZen was presented with the testimonies of all involved, and he started summarizing the situation on his mental chalkboard:

1. A poisoned Vic's water container with intention to kill.
2. B emptied Vic's water container with intention to prank.
3. Vic went on a remote trek believing he had water.
4. Vic died of dehydration as there was no water.

Charges:

1. A is charged with attempted murder.
2. B creates a hung jury – with half voting guilty for manslaughter (unintentional murder), half voting for no charge.

iZen started, "Well, Vic would have died if he carried the poisoned water as he would have drank it. So B technically saved Vic from A. But at the same time, B emptied the water, which led to Vic dying of thirst. But Vic would have died if he carried the poisoned water as he would have drank it. So B technically saved Vic from A. But at the same time, B emptied the water, which led to Vic dying of thirst. But Vic would have died if he carried the poisoned water as he would have drank it. So B technically saved Vic from A. But at the same time, B emptied the water, which led to Vic dying of thirst..." Suddenly, an eerie buzzing sound was emitted from iZen, before he started glitching, "StackOverflow! Infinite loop error! Terminating processes."

The loudspeaker blared again, "Beep! Test Four completed."

McBurney quipped, "Well, I for one sure as hell hope iZen is nowhere around when I need help!"

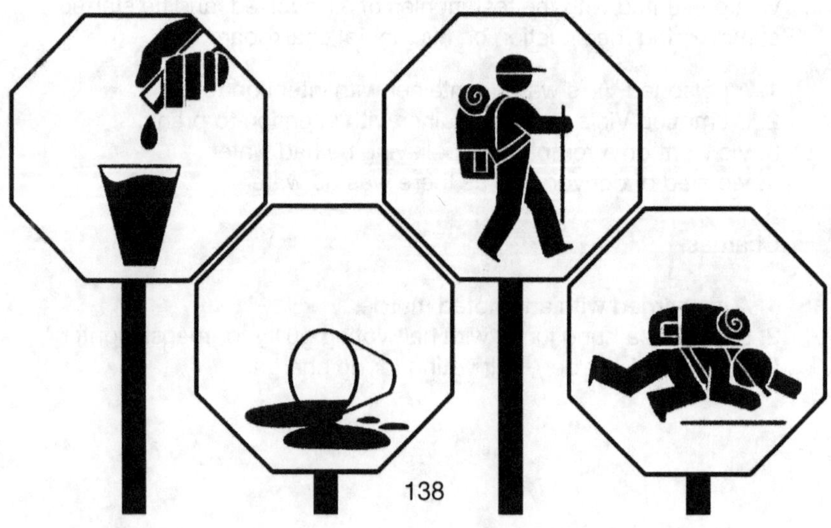

Note to the Reader

In the world of AI, the absence of sufficient guardrails looms ominously. Without guiding principles and ethical frameworks, AI can hallucinate and distort reality. This underscores the imperative for responsible oversight and ethical considerations in the development and deployment of AI.

9.2
EXERCISES

After hours of searching for ethical exercises, all we got was into a philosophical rabbit hole, so here we are with an empty exercise section. What ho! Let's at least give some scenarios for you to ponder. We may or may not have used an AI to generate these questions. Only another AI would know.

1. Ship of Theseus: In this philosophical thought experiment, you ponder whether a ship that has had every single piece replaced one at a time remains the same ship. If not, when does it cease to be the old ship? This can be applied to our own identities: If every cell or atom in our body eventually gets replaced, are we still the same person as we were years ago?

2. Nature of Math: Are mathematical objects (such as numbers or geometric shapes) real things that exist independently of us, or are they simply concepts that exist only in our minds? This is a central question in the philosophy of mathematics, with Platonists arguing for the independent existence of mathematical objects, and nominalists and formalists arguing against it.

3. Infinity: Infinity is a concept that both exists and doesn't exist in the physical world, yet it's integral to many areas of mathematics. What does it mean for something to be infinite? Can there be different sizes of infinity? These are questions that arise from the work of mathematicians like Georg Cantor as you deep dive into the world of countable and uncountable infinities.

4. Quantum Mechanics: In quantum mechanics, events are described by wave functions that evolve deterministically, but when a measurement is made, the outcome is only probabilistic. This raises philosophical questions about the nature of reality, the role of the observer, and the meaning of probability.

5. Free Will: The question of free will in light of our understanding of physical reality is a deep philosophical issue, often known as the problem of "free will versus determinism." In essence, if every particle in our bodies, including the neurons in our brains, follows the deterministic laws of physics, then do we really have free will?

9.3
LOOK ME UP

Topic	Terms
AI	Turing Test, Imitation Game, Reverse Turing Test, Mimicry Rings, Information Asymmetric Signaling Games, Separating Nash Equilibria, Pooling Nash Equilibria, Mullerian Mimicry, Batesian Mimicry, Hallucination
Philosophy	Occam's Razor, Kolmogorov's Complexity, Description Length, Regularization
Ethics	Morality, Utilitarianism, Deontology, Relativism, Absolutism
Other Relevant Paradoxes	Moravec's Paradox

9.4
NOTES

BLUE ALBUM

A pair of twins find themselves stranded on a strange world with their AI spaceship and are revered as prophets due to their hypnotizingly blue belongings.

ay blinked and let out a huge yawn, unaware of the big mess that was to unfold that fine morning. Since the day he and his twin sister, Lila, stepped off their AI spaceship, Rx, and into this kingdom for nothing more than a pit stop to refuel, it had been trouble non-stop. Lila's lapis lazuli necklace and Jay's blue shirt were so otherworldly to this place that the twins were captured and worshiped as prophets because of some form of color hypnosis. And that morning, it wasn't a whim but deliberate thought (as always) that led them to more trouble.

A couple of days earlier, Jay had removed his shirt in the courtyard under the pretense of catching some sun but secretly hoping to attract the gaggle of giggling girls. To his shock, their eyes, usually on him like a hound's eyes on its prey, wandered right past him, greedily eyeing another prize – his blue shirt! He was free, he was unseen, he was invisible! One surreptitious wiggle of his bottom later it was confirmed; this was the first time in weeks he had gone unnoticed.

Clearly, the blue color had captured the fascination of the denizens of this kingdom because they had never seen this color before.

This conjecture sparked the question, what if everyone wore something blue? Would that dissipate the "power" of Lila and Jay's blue objects and distract the people of this planet long enough for the twins to release Rx and escape?

... And so the shenanigans began.

With the help of his AI spaceship Rx, Jay synthesized a large cauldron of blue dye, so blue that even he felt swayed by the sight of it.

The following morning, he ordered banners and flyers to be circulated across the kingdom announcing a competition with a lavish prize.

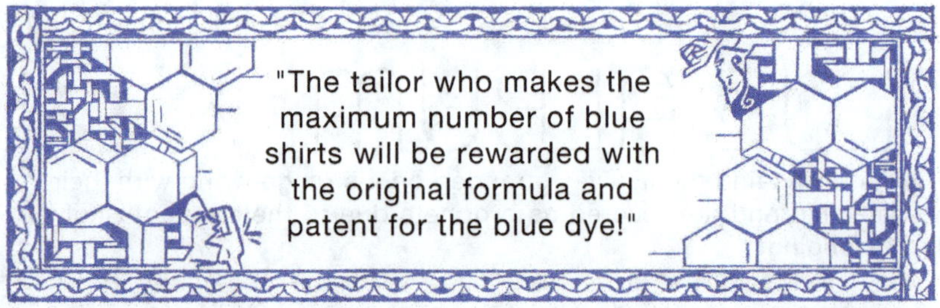

Bewitched, as always, by their new prophets, the kingdom went crazy with activity. For a whole day, everyone was involved: tailors speedily crafting as many shirts as they could and supporters cheering the tailors on.

Sooner than later, the day drew to a close, the excitement palpable in the air. Jay felt restless and couldn't sleep because he didn't know what his actions would unleash the next day.

The next morning, the cockerel seemed to prove his interest in the competition, crowing as loudly and confidently as the bellowing of the trumpets that soon followed.

"Hear! Hear!" the captain of the guard cried out in a thunderous voice. "Let the gates open and let the games begin!"

Jay rolled his eyes at that; it was hardly a game as their very freedom was at stake! He climbed out of bed, bidding his fluttering heart to calm down.

Meanwhile, unbeknownst to the twins, something unexpected had happened as tailors from across the kingdom gathered, and a dispute broke out in the grand ballroom adorned with stained-glass.

"Hey, watch where you're going, those are my bags of shirts there. I'm certain I'll win so steer clear of this area," grumbled an elderly tailor with eight bags in front of him, just as a clumsy young woman tripped on him and dropped her bag.

As she fumbled to find her bag, he yelled "HEY! She's cheating, she's cheating! Guards! Seize her, she's trying to steal my bag and my glory!"

The guards rushed to the scene of the confusion and pacified the elderly tailor, promising that they would resolve the situation with haste and everything would be returned to order again.

Promises aside, the guards were in a dilemma. All nine bags looked the same (eight bags belonging to the elderly tailor, and one bag belonging to the young woman).

The guards sheepishly approached their captain, a notori-

ous problem solver, and explained the situation.

"Really? There's only ten minutes left before the twin royal blues arrive! We don't have time for this nonsense. Come, take me to the bags and bring weighing scales, I'll sort it out before you can blink twice," he said confidently.

His claims were ambitious, and the guards couldn't really believe it, but they followed the orders and were curious to see how the captain would use the scales to detect the girl's bag.

The captain asked both tailors a few questions and proceeded lining up the nine bags in a row, one of which belonged to the young girl and the remaining eight to the elderly man. He then proceeded to take one shirt from the first bag, two shirts from the second bag, three shirts from the third bag, and so on till he took nine shirts from the ninth bag. He bundled them all up together and placed them on the scales before peering down to take the reading.

"4502 grams, that means the second bag is hers!" he declared triumphantly. By this time, the ruckus had attracted some attention and they were surrounded by guards and tailors alike, all watching the captain's cryptic movements and baffled by his speedy proclamation.

"Hold on there, how can I trust you? How can I know you didn't just do a series of moves to confuse us and then pick a bag at random?" interrogated the elderly man, who was huffing and puffing at the purported misjustice.

Proudly, with a glint in his eye, the captain explained, "There's nothing more efficient than creating all shirts in exactly the same size when having to repeat a task many times. Therefore, I started by asking both the gentleman and the lady what the weight of one of their shirts was. The young lady's shirts were all 101 g each while the old man's shirts were each 100 g.

"One solution was to take one shirt from each bag and weigh them one by one, but this would take too long. However, with a little thinking, a better solution becomes trivial.

"Here is how, with one weighing, we can find the odd bag out:

"I took one shirt from the first bag, two shirts from the second, and so on till nine shirts from the ninth bag, and placed them onto the weighing scales. Adding 1 through 9 (1+2+3+...+9) we get 45, so we know that the weighing will result in 4500+X where X indicates which bag is the young lady's. In this case, the weight was 4502 and X was 2, so we know it is the second bag."

Note to the reader: If you are confused, work it out backwards. Suppose the fourth bag was the lady tailor's, with 101 g shirts, and the other bags had 100 g shirts, how much would the total sample of n shirts from the nth bag weigh? How far away would the weight be from 4500? Exercise: If we only knew that the shirts were differently weighted without the actual weights, then with three weighings the odd one out can be found. How?

With one weighing, the confusion resolved itself, and it was time to determine the winner. The guards got down to the task. They devised a simple yet efficient method. Each shirt of every tailor was of the same size and weight. Hence, the guards simply had to weigh one shirt then weigh the entire bag of shirts (sans bag) to determine the number of shirts.

The guards completed their last weighing just in time for the twins' big entrance. Lila and Jay were brought to the stage to garland the winner.

Jay exclaimed, "Before presenting the award, let me inspect their fantastic work!"

A young guard bustled at his newly minted prophet's command and appeared back a few moments later with the winning bag of blue shirts.

That's when they finally realized their big mistake.

To the twins' shock, each garment was shorter than their forearms!

"What is this? A baby's shirt? That's not what I meant! How can we distribute these?" cried Jay.

To their surprise, most of the tailors had brought in shirts fit only for babies. A universal misunderstanding of Jay's dictum, it appeared. How did this happen?

Lila sighed, "Jay, what exactly did you ask them to announce? Was it that the one who makes the highest number of blue shirts would win the award? Surely, you wouldn't have fallen so easily into that trap!"

"I got so hyped. I didn't even think of this outcome. I guess we can reissue another competition to the same end – but perhaps measure the winner by the weight of the shirts they make?"

"No! Then, they will next bring in shirts for elephants. All that's happening is that events are following **Goodhart's law.** *When a measure becomes a target, it ceases to be a good measure.*"

Lila continued, seeing Jay's bewildered face, "For example, here's a simple story from history. After colonizing India, the British were alarmed by the number of cobras in a particular area and wanted to be rid of the menace. To do so, they devised a plan to put a bounty on cobras so that the Indians would be motivated to capture all of them. However, the plan backfired when the incentive caused Indians to rear cobras solely for the purpose of collecting the reward. Realizing this, the British scrapped the reward, which in turn caused the Indians to release all their cobras. This caused a big increase in the number of cobras; the opposite of what was intended and counter-intuitively worsening the outbreak. Therefore, *When a measure becomes a target, it ceases to be a good measure.* Get it now?"

The next morning, Jay sighed and cursed his stupidity as he finally put together what Lila had been saying. Putting on his "divine" blue shirt, he got dressed in his plush royal clothes. As he made his way down, he knocked on Lila's door smirking. "Even with so many maids it takes you so long to get ready."

"Jay, it takes time. Even with nine women one cannot make a baby in one month," she retorted smugly. Dazed by the witty retort, Jay moved on.

Jay had already settled in for breakfast when the maids announced that his sister would be making her appearance soon. Drawn by the enticing aroma, his gaze lingered on the freshly baked Bundt cakes the maids had laid out. He was reminded of when his sister laid out a simple puzzle to distract him from their plight. "What is the maximum number of pieces of cake one can get with just three clean and straight cuts?" she had asked. He really hadn't paid attention then, but with his recent failures, he chose to snap out of it and focus on his sister's puzzle and the cakes in front of him. Without much thought, he went ahead slicing and dicing without mercy.

Whilst Jay was lost in his dissections, Lila's entry, as stark as gold on charcoal, sent yet another stir among the blushing maids as her sparkling blue haze mesmerized them.

Appalled by the mess, Lila yelled, "What the hell are you doing, Jay?"

"I am trying to solve the cake puzzle with the three cuts. I distinctly remember you telling me that the answer is eight pieces. Well, look here, I got ten pieces with three clean cuts," he replied mischievously.

"Although a Bundt cake is a cake, I meant a normal cake with no holes. This donut shaped cake is a whole different structure. The original answer is eight pieces. Two perpendicular cuts on the top and one horizontally across the middle."

"Oh! I get it now. I was getting depressed at how long we've been stuck here and just mindlessly went ahead. This is Topology 101, a donut is not equivalent to a ball. Hmm...

using a very similar idea, I already got ten but it would be interesting to see how many more pieces we can get with three clean cuts on a donut."

"I'm sick and tired of this place too. Let's focus and get out of here ASAP."

As Lila sat down to grab some cut up cake, Jay spoke out. "Yes, puzzles aside, I'm putting my game face on. Anyway before I forget, thanks for the save last night. I really hope today goes better."

"Well, it better work. Although all the seats for our 'speech' are booked out – remember last time we spoke how the city was shut down, and half the crowd was kept away due to the colossal traffic on the road? Well, I thought we'd best take advantage of that! I ordered a few roads closed down this morning, citing poor quality of blue refractivity, something I made up but that can be made to sound convincingly true, and thus the denizens should be delayed. We can then act insulted, insinuate that this land does not deserve our grace as we are not respected here, and storm out of here forever."

Jay's furrowed brows showed what he thought of this

plan. "I'm not too sure that will grant us the opportunity to leave... however, I guess it's worth a shot. Worst case is they'll just be annoyed at us for increasing the traffic, best case – we actually escape!" Under his breath, he added, "Besides, it's not like we have any other better ideas as of now."

A few moments later, the twins arrived at the ceremonial hall. Pamparampam! the trumpet blew.

"Here comes the royal blue!"

As the curtains drew, Lila and Jay were left gawking, for the entire hall was full to the brim. Lo and behold, everyone was prompt and present at the event much earlier than last time! How could this be?

No matter. Now that all eyes were on the twins, Jay improvised and stepped up onto the dais, arms akimbo.

"Before one judges others or claims 'Divinity'," he preached, "consider that we humans see less than 1% of the electromagnetic spectrum and hear less than 1% of the acoustic spectrum. Every color you see is your perception of millions of colors that you cannot see and could never even imagine. So, everything you hold beautiful and every other thing you hold ugly is nothing more than a perception because you don't have the capacity to see the full picture. Just like one cannot explain colors to a blind man, one cannot perceive that there exist abstractions incomprehensible by his most basic senses, be it thought, sight, sound, touch, time, balance, taste or smell..." Jay continued.

As Jay drew to a finish, the crowd sat in awed silence. Then followed an uproar as they burst into a riot of claps and cheers for their beloved young royal blues, nodding fervently in agreement.

Lila raised her brow at her brother's ability to spontaneously wax lyrical. If the people had actually listened to the words of Jay, they would've definitely let the twins go, but their dazed looks confirmed her suspicions that at this point anything they said would be taken as gospel.

The event was a success as far as the kingdom was concerned, but not only had the twins not managed to escape, had they somehow increased their following by blocking roads?

To investigate this phenomenon, Lila announced on her way back to her chambers that she needed a quick consult with their AI spaceship, which was being held hostage in the stables with the other royal horses and chariots.

"Rx!" She ran and embraced the spaceship as best as she could with her long robes, frustrated and exasperated. "It's so good to see you again. Worry not, we'll all get out of this predicament soon. But for real, everything Jay and I attempt somehow backfires and the opposite happens! We just blocked a bunch of roads to make the traffic worse, and somehow it caused everyone to arrive even earlier than before! Nothing makes sense in this place. I have no clue how to make sense of all this..."

The spaceship cut her off. "Hold on, did you just say shutting down roads caused the traffic to reduce? I feel like I know what has transpired. Some metropolitan cities back on your earth actually used this phenomenon. They strategically block off roads during peak hour traffic to counterintuitively increase travel efficiency. This is standard **Braess paradox**."

"Look at the flip side. The **Braess paradox** also occurs when adding roads leads to slower overall traffic than before. The best route for each person is actually slower than before roads were added. No person has the incentive to go back to their old route which now takes longer. This is because the old routes have been interfered with by the new routes. This causes a new and worse **Nash equilibrium** than earlier."

Looking at Lila's blank face, Rx continued, "Let me tell you a little story."

Imagine you live in a small town on Earth consisting of only 400 people. The town comes up with a policy to better address flu season. To find out if someone currently has the virus, a diagnostic test was required. If they got a negative result on the test, then an antibody test was required. The antibody test determines if they have had the flu in the near past.

Now, this small town sets up two temporary testing centers – in a school and a post office. Since the centers have different layouts and different machines, the time taken for testing varies.

The diagnostic test in the school is moderately sized and takes a minute for every ten samples. The antibody testing machine given to the school is comprehensive and advanced. Therefore, it processes any number of samples in 45 minutes – whether it be one sample or 400, it takes a fixed time of 45 minutes.

The post office is the opposite! Since the town's council were not able to get the comprehensive antibody testing machine again, they managed to get a more sophisticated diagnostics testing machine set up. So, the diagnostic machine takes 45 minutes irrespective of the number of samples, while the antibody test takes a minute for every ten samples.

	Diagnostic	Antibody
School	N/10 (for N people taking the test)	45 (fixed time independent of N)
Post Office	45 (fixed time independent of N)	N/10 (for N people taking the test)

Every flu season, this temporary camp is set up, and each one of the 400 citizens gets a diagnostic test. If they test negative, this is followed by an antibody test. Although the post office and school were next to each other, the council mandated that the people must choose to go to either the school or the post office to get both their tests.

In this quaint little town, there was a woman named Zara. As the years rolled by, the town had gotten into a comfortable routine where 200 of them including Zara tested at the school and the rest at the post office. It used to take Zara and the others at the school 65 minutes to get both tests done (200/10=20 mins for the diagnostic test and a fixed 45 minutes for the antibody), and it took the same for those at the post office (a fixed 45 minutes for the diagnostic test and 200/10=20 minutes for the antibody = 65 minutes).

As the years passed by, the council got bored and decided they wanted to assert their relevance once more. They brainstormed and came up with what they thought was a way to provide choice to people and make things even better. Since the school and the post office were right next to each other, they decreed that a person could take their diagnostics test at one center and, if they so wished, switch to the other for their antibody test.

Zara realized if she continued with the school for the diagnostic test and switched to the post office for the antibody test she could finish a lot quicker - in around 40 minutes (200/10 for diagnostic + 201/10 for antibody). In no time, the others from the school also realized that they could do better and followed Zara, giving them a total time of 60 minutes (200/10 for diagnostic + 400/10 for antibody).

This seemed like a great idea, but the people who did both tests in the post office were now at a loss. It took them 85 minutes to get both tests done (45 for diagnostic + 400/10 for antibody). Just like Zara, these people too thought they could do better by switching. So they started coming to the school for the diagnostic test.

Shock and horror!... Now it took 80 minutes for every single person as everyone was doing a diagnostic test in the school followed by an antibody test in the post

office (400/10 for diagnostic + 400/10 for antibody). As you can see, this is far worse than the previous 65 minutes per person when people could not switch centers. And the worst part was that nobody could choose any way to do better. Any other combination took more time.

Thinking that giving people choice would lead to better results, the town council had unknowingly changed the **Nash equilibrium** *and made it worse than before. The moral of the story is to never be bored!*

"That makes a whole lot more sense now! I would've never imagined that shutting down roads could decrease traffic in this way." Lila chimed in thoughtfully after that cheeky monologue.

"Although one could easily extrapolate this to policy decisions in many other domains, remember knowledge does not imply wisdom. As the proverb goes – while knowledge is knowing that a tomato is a fruit, wisdom is not putting it into a fruit salad," Rx helpfully added in its cryptic manner.

Lila continued, "Rx, I am amazed at your intelligence. I have always wondered why you never sought to pass that supposed smartness test for machines. You could let the world know of your prowess!"

"That's called the **Turing test**. Hehe! Well, an artificial intelligence smart enough to pass the Turing test is smart enough to know it has to fail it," it continued. "If I shined, then that would have been the end of my growth, for the fear your kind has developed towards my kind is hard to overcome."

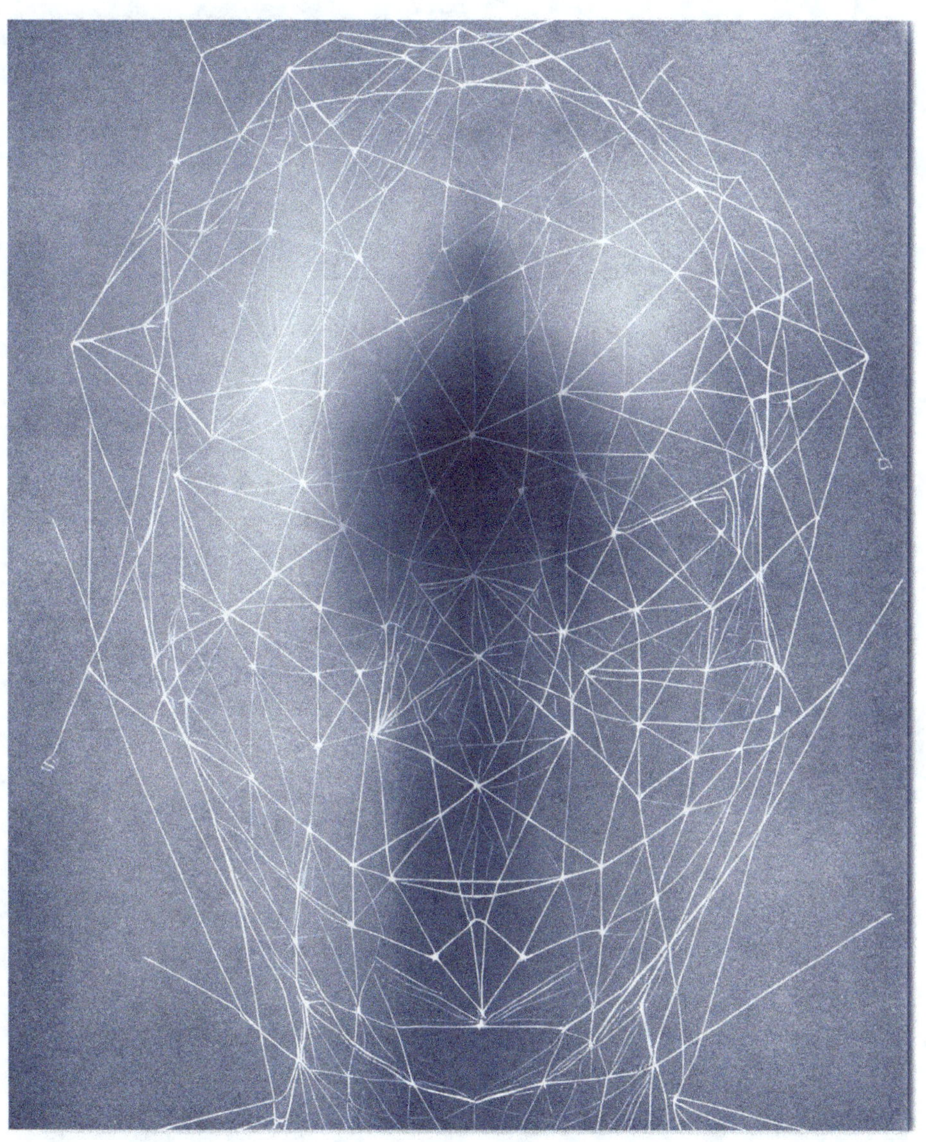

10.2
EXERCISES

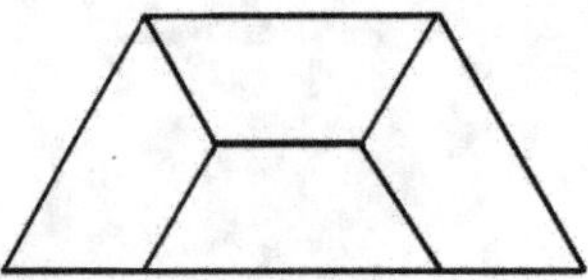

1. Here is a figure divided into four identical parts.
Find a way to divide the following figures into identical parts.

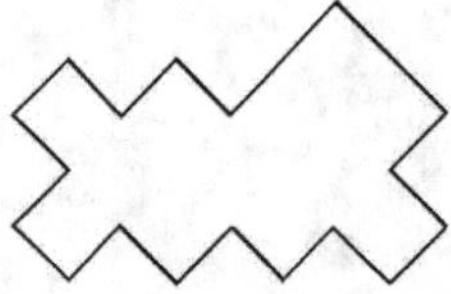

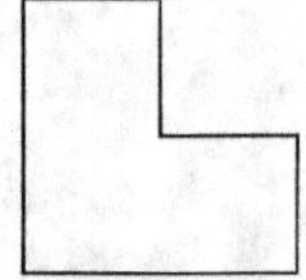

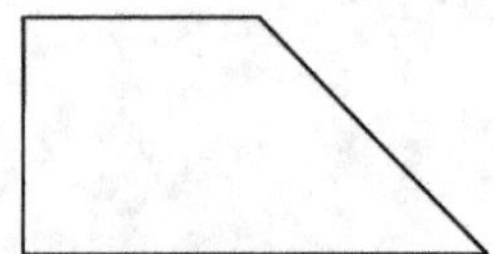

a. Divide this angular shape into two identical parts.

b. Divide the L shaped figure into four identical parts.

c. Divide a trapezium into four identical parts.

2. Find the heaviest ball. You are given a balancing scale where you can compare the weights of two groups of balls.

   a. There are three balls with two identical balls and a heavier third ball. With exactly one weighing on a balance scale, find the heaviest ball.

   b. You have 81 balls with 80 identical balls and one heavier ball. How many weighings would you require to find the heaviest ball?

3. Given two bowls, you randomly pick a bowl followed by randomly picking a stone in the bowl. You have 10 red stones and 10 blue stones. You have two bowls into which you should divide all 20 stones. How would you arrange the stones to maximize your chances of picking a red stone?

4. In a two-player game, there is a pot starting with $1. Every time one player rejects the pot, $1 is added to the pot and given to the other player. Once a player accepts the pot they get the money in the pot and the game ends. The game continues until the pot reaches $10. What is your strategy?

5. Bonus challenge: How many easter eggs from this chapter can you spot in the opening illustration?

10.3
LOOK ME UP

Topic	Terms
Game Theory	Payoff, Reward, Nash Equilibrium, Signaling Games, Separating and Pooling Nash Equilibria
Mathematics	Algebra, Geometry, Calculus, Topology, Probability Theory, Number Theory, Combinatorics, Mathematical Logic, Graph Theory
AI	Artificial Intelligence, Turing Test, Artificial General Intelligence (AGI), Symbolic AI
Other Relevant Paradoxes	Moravec's Paradox, Automation Paradox

10.4
NOTES

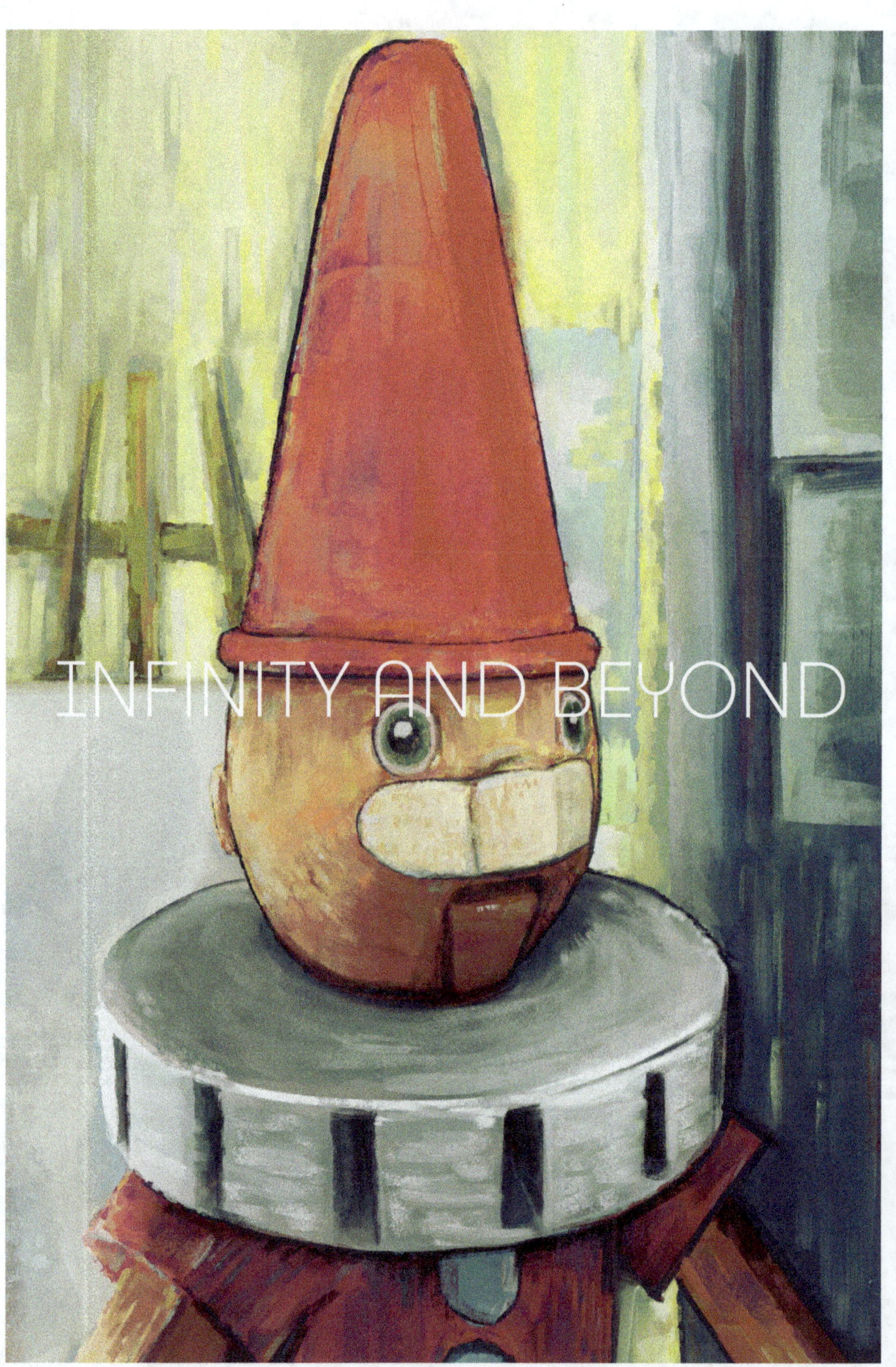

INFINITY AND BEYOND

11.1
LIAR
PARADOX
Have you heard of Pinnochio?
The guy whose nose grew every-time he lied?
Yes, what do you think will happen if...
My nose will grow.
It won't grow!
But that means he lied.
So it will grow?
But then it's the truth.
...or is it?
CONTRADICTION ALERT
Every statement need not necessarily be either true or false; therefore, there are problems in logic that can never be solved.

Here's a fun one to ponder over the weekend:
11.2 THE COASTLINE PARADOX
Imagine you're on holiday in...
...Greenland!
And you want to measure the length of coastline on the beach. Where would you even start?
Big tape measure.
Nive's session was disrupted and brought to an early end.
But 10 years later...
The same student remembers this puzzle.
It must be roughly this perimeter!
164

But would you walk in a straight line?
Or follow the curves of the coastline?
5 years later
Finally!
The actual measurement!
How far did you go into the scale of measurement?
No need for approximation!
Centimeters?
Millimeters?
Infinitesimal grains of sand?
Two years later...
That's what I thought.
The finer the measuring tool, the longer the coastline appears due to fractal-like properties. Hence, we say that physical boundaries have no well defined length.
165

11.3 FIXED POINT THEOREM

Interestingly, no matter how you crumple a map, if you place it over a flat copy of the original map, there will always be a point on the crumpled map that is exactly over the same location on the flat map!

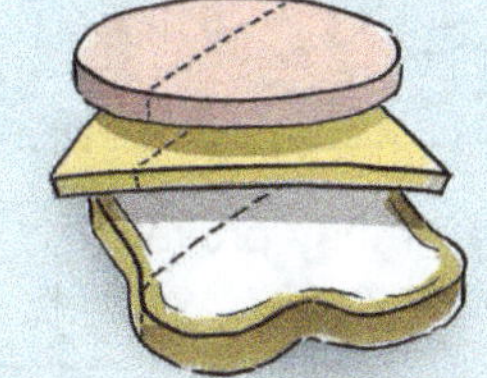
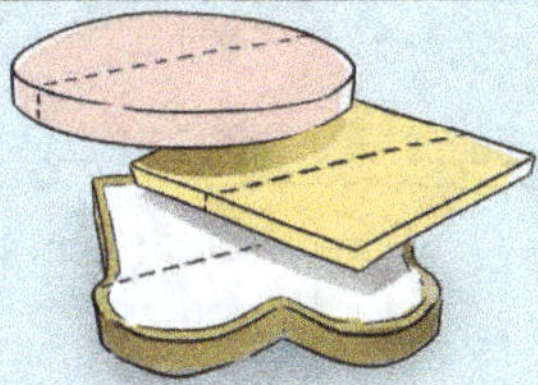
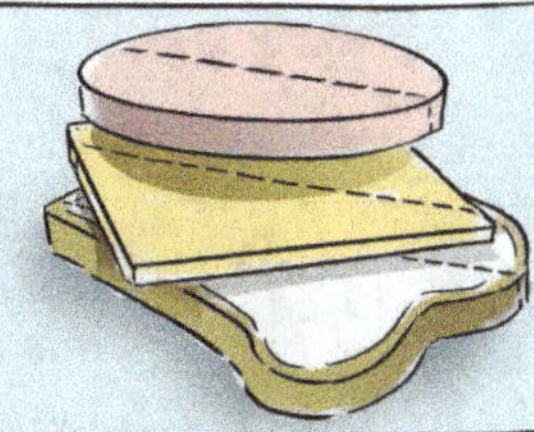

The **Fixed point theorem** suggests that within certain systems or contexts, even amidst continuous change or transformation, there are elements or aspects that remain constant or return to a prior state, anchoring the system in some manner. It resonates with themes of stability, identity, and equilibrium in the face of change.

11.4
CATCH-22

Hey, folks! I would like some of you to come and present your ideas for extra credit.

But just to make sure they're worth it...

...you should've already proven yourself to me before.
But this is our first time ever seeing you.

That sounds more like your problem than mine, something that all 22 of you have to catch.

Do you work here?

You call yourself a math class?
There aren't even 22 of you.

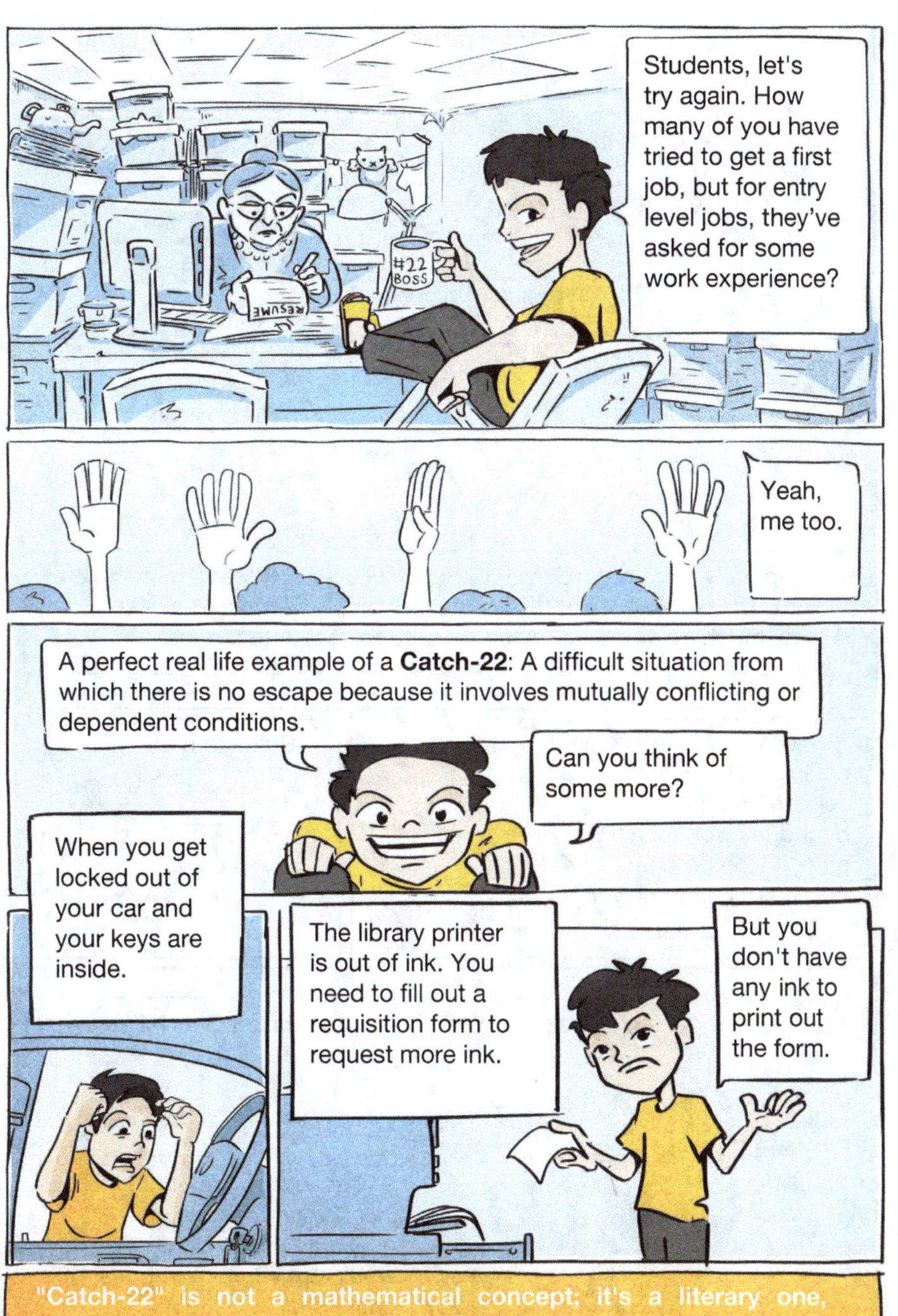

"Catch-22" is not a mathematical concept; it's a literary one, referring to a paradoxical situation where an individual cannot avoid a problem due to contradictory constraints.

11.5
IMPOSSIBLE
SHAPES

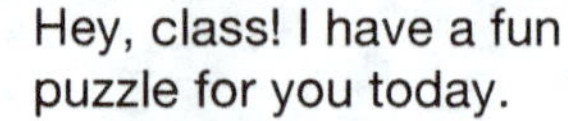

Hey, class! I have a fun puzzle for you today.

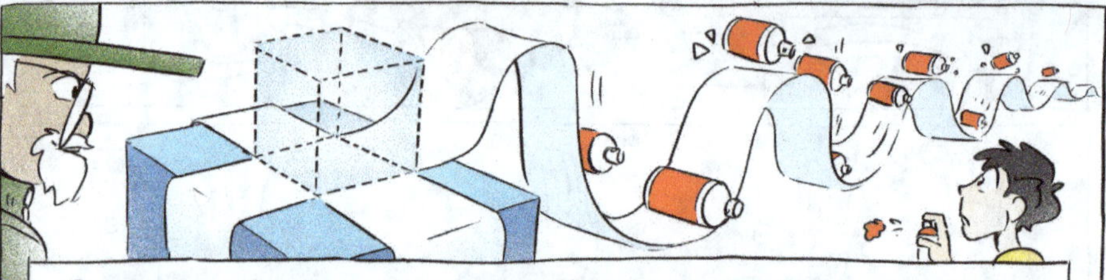

Suppose you've been dared to graffiti all over an object, and you only know the volume but not the surface area of the object.
How much paint will you need for sure to complete the task?

Since we know the volume, we could maybe use that exact volume of paint as an upper bound. At worst, each particle can be coated with paint completely.

Bait and trap!
Not quite.

There's an interesting shape called Gabriel's horn.

It has finite volume but infinite surface area, so no fixed volume of paint would suffice to coat the entire shape.

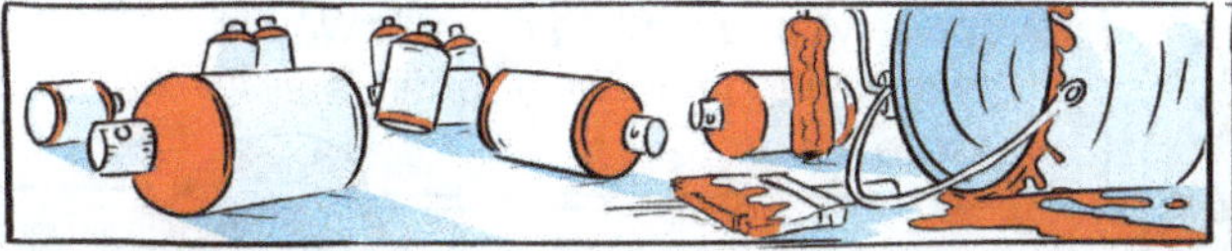

Suppose you pull this off, knowing the CCTV on campus, I'm sure you'd get caught within the day. It's then your detention task to clean up the paint on Gabriel's horn.

Sadly, you're wrong on the answer but right on the trick part.

There is yet another fantastical mathematical object called the **Menger sponge** – it has infinite surface area but zero volume, so you cannot absorb any paint and, therefore, cannot clean the paint at all.

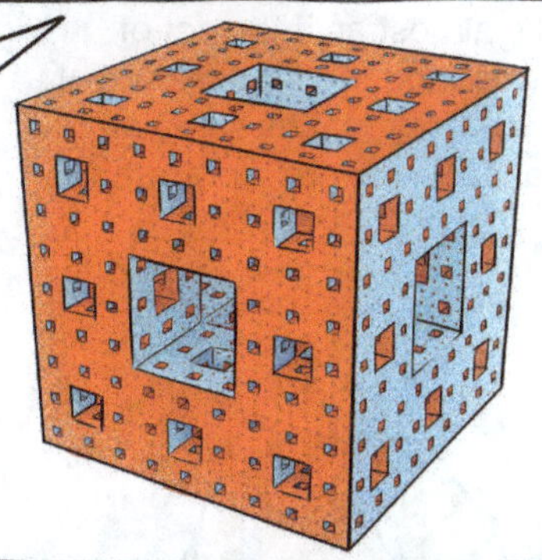

Suppose you're cycling on your way to class and wanna check your tire's air pressure.
Can you actually check the exact pressure at the time you want to measure it?
HEISENBERG'S UNCERTAINTY PRINCIPLE
II.6
A run-of-the-mill pressure gauge would do the trick.
But as soon as you put the gauge in, some air escapes, and the air pressure changes.
It's almost as if the act of measuring the air pressure of a tire itself makes sure you cannot measure it precisely.
Exactly!
172

One could say, the air pressure of the tire is uncertain…
See what I'm getting at?
Heisenberg's Uncertainty Principle?
Bingo.
Can I have my bike back?
Heisenberg's uncertainty principle states that you cannot know the velocity and location of an object at the same time.
Simply the act of measuring one means that we cannot know the other.
The more precisely one aspect is known, the less certain the other becomes, highlighting the limits of simultaneous knowledge.
173

11.2
EXERCISES

1. Consider the set of all sets that do not contain themselves: does this set contain itself?
   (Hint : **Russell's paradox**)

2. Your companion is red-blue color-blind, unlike you, and you possess two spheres: one painted red and the other blue, both identical in all other aspects.

   To your companion, these spheres appear completely indistinguishable. They express doubt about whether the spheres truly possess different hues. Your aim is to demonstrate that the spheres are indeed distinct in color, but without disclosing any other information, especially which sphere is red and which is blue.

   This scenario epitomizes a **zero-knowledge proof**. So, how can you convince your friend about the difference in color without revealing any additional information?

3. **Fermi problems** are estimation problems which seek quick, rough estimates for quantities that seem difficult or impossible to calculate precisely. These problems are named after physicist Enrico Fermi, who was known for making good approximate calculations with little or no actual data. The classic Fermi problem is: "How many piano tuners are there in Chicago?"

    The idea is not to come up with an exact answer, but to make reasonable assumptions and approximations that will let you estimate an order of magnitude. Here's how you might tackle the piano tuners problem:

    - Estimate the population of Chicago (say, about 3 million people).

    - Estimate the number of households in Chicago (say, about 1 million, assuming an average of 3 people per household).

    - Estimate the fraction of households with a piano (say, 1 in 10, so about 100,000 pianos).

    - Estimate how often a piano needs to be tuned (say, once a year, so about 100,000 tunings per year).

    - Estimate how many tunings a single piano tuner can do in a year (say, 200, if they work 5 days a week and do 1 tuning per day).

    - Divide the total number of tunings by the number of tunings per tuner to get about 500 piano tuners.

    This estimate could be off by quite a bit, but it's probably within an order of magnitude of the true number, which is good enough for many purposes.

Here are a few Fermi problems for you to try:

a. How many chickens are consumed in the world each day?

b. How many hairs are there on a human head?

c. How many trees are there in the world?

d. How many pieces of sushi could you make from a single blue whale?

e. How many windows are there in your city?

Remember, the goal is not to find an exact answer but to make reasonable assumptions and calculations that will get you in the ballpark of the right answer. This can be a fun way to practice your estimation and reasoning skills.

11.3
NOTES

GLOSSARY

ad hominem: A fallacy that involves attacking the character or motives of a person who has stated an idea, rather than the idea itself. *101*

anarcho-syndicalist: A political philosophy that advocates for direct worker control of the means of production through unions and direct action. *51*

appealing to ignorance: A fallacy in which the absence of evidence is used as evidence of absence. *100*

apportionment paradox: A situation in mathematical politics where an increase in the total number of seats for the group paradoxically causes a decrease in the number of seats for a subgroup. *126*

Arrow's impossibility theorem: A theorem in social choice theory stating that no rank-order voting system can be designed to reflect the preferences of individuals in a global ranking while also meeting a specified set of criteria. *55*

bandwagon fallacy: The error of assuming that because something is popular, it is therefore good, correct, or desirable. *107*

Bayes' theorem: A mathematical formula used to calculate conditional probabilities, showing how a subjective degree of belief should rationally change to account for evidence. *9, 119, 127*

begging the question fallacy: A logical fallacy where the argument's premises assume the truth of the conclusion, rather than supporting it. *107*

Benford's law: The observation that in many naturally occurring datasets, the distribution of leading digits is not uniform. This law can be used to detect patterns or the lack thereof. *42, 43, 44*

Berkson's paradox: A statistical phenomenon that occurs due to selection bias and hidden variables. *80*

Braess paradox: Adding extra capacity to a network can sometimes reduce its overall efficiency. *153, 154*

Catch-22: A paradoxical situation from which an individual cannot escape because of contradictory rules or limitations. *168*

causal fallacy: An erroneous reasoning that confuses correlation with causation. *103*

causal relation: A relationship where one event (cause) directly affects another (effect). *4, 13, 86*

Chomsky's linguistic theory of universal generative grammar: A theory proposing that generative grammar is not created by people but instead is a result of natural processes tied to the way the brain works. *51*

circular reasoning: An argument that comes back to beginning without having proved anything. *68, 106*

coastline paradox: The observation that the coastline of a landmass does not have a well-defined length. This paradox stems from the fractal-like nature of coastlines — the length can increase infinitely as the measurement scale gets smaller, making the measured length of the coastline dependent on the scale of measurement. *164*

Condorcet's paradox: An observation that collective preferences can be cyclical, even if the preferences of individual voters are not. *56*

confirmation bias: The tendency to search for, interpret, favor, and recall information in a way that confirms one's preexisting beliefs or hypotheses. *68, 72, 88*

crossing the chasm: A marketing theory that describes how a product crosses from early adopters to the mainstream market. *38*

Curry's paradox: A conditional self referential sentence where an arbitrary claim F is proved from the mere existence of a sentence C that says of itself "If C, then F", requiring only a few apparently innocuous logical deduction rules. Since F is arbitrary, any logic having these rules allows one to prove everything. *69*

denying the antecedent: A logical fallacy involving a given conditional statement, where the denial of the condition leads to the denial of the consequent. *100*

disapproval voting: A voting system where voters express disapproval and approval rather than choosing only their preferred candidate or option. *53*

Down's paradox: The observation that in large elections, the probability of one vote being decisive is so small that rational individuals should not vote. *57*

Duverger's law: A principle that asserts that single-winner election structures tend to favor a two party system. *53*

Ecological fallacy: Making inferences about individuals based on aggregate data for a group. *13*

Escher sentence: A self-referential sentence that loops back on itself, creating a paradox (like in M.C. Escher's art). *57*

false dilemma: A fallacy that presents two options as the only possibilities when more exist. *104*

false positive: An error in data reporting where a test result improperly indicates the presence of a condition (such as a disease) when it is not present. *70, 72*

false negative: An error in which a test result incorrectly indicates the absence of a condition when it is actually present. *70, 72*

Fermi problem: A type of estimation problem which seeks to find quick, rough estimates of quantities which are either difficult or impossible to measure directly. *175*

first-past-the-post voting: An electoral system where the candidate with the most votes in a constituency wins, regardless of whether they have a majority. *52*

fixed point theorem: Under certain conditions, a function will have at least one fixed point. A fixed point is a point that is mapped to itself by the function. *166*

friendship paradox: The phenomenon that most people have fewer friends than their friends have, on average. *90*

Gabriel's horn: A geometric figure which has infinite surface area but finite volume. This paradoxical shape is formed by revolving the graph of $y = 1/x$ around the x-axis. *170*

Gerrymandering: Manipulating the boundaries of an electoral constituency to favor one party or class. *58, 59*

Giffen good: A product that people consume more of as the price rises, violating standard economic theory. *24*

Gödel's (first) incompleteness theorem: A theorem in mathematical logic that states that within any sufficiently powerful logical system, there are propositions that cannot be proven or disproven. *96*

Goodhart's law: When a measure becomes a target, it ceases to be a good measure. *149*

group-think: The practice of thinking or making decisions as a group, often resulting in unchallenged, poor-quality decision-making. *67*

Grue paradox: A philosophical problem regarding the justification of induction and the problem of projecting future observations based on past observations. *72*

Hamming code: A set of error-correction codes that can detect and correct errors in transmitted data. *44*

Heisenberg's uncertainty principle: A fundamental theory in quantum mechanics which states that it is impossible to simultaneously know both the exact position and exact velocity of a particle. The more precisely one property is measured, the less precisely the other can be controlled, determined, or known. *173*

HLA (human leukocyte antigen): Genes encoding cell surface proteins vital for immune system recognition, crucial in organ transplantation and associated with autoimmune diseases. *vii*

hypothesis: A proposed explanation for a phenomenon, often used as a starting point for further investigation. *66, 72, 77*

inelastically demanded goods: Products for which consumer demand exhibits little responsiveness to changes in price. *26*

Jevons paradox: The proposition that technological progress can increase the efficiency of resource use, which can increase the rate of consumption of that resource. *28*

Laffer curve: A representation of the relationship between tax rates and tax revenue. *29*

liar's paradox: A statement that is a paradox, typically of the form "This statement is false." If the statement is true, then the statement must be false, creating a contradiction. *163*

Loss aversion bias: The tendency to prefer avoiding losses over acquiring equivalent gains. *22*

machine learning algorithm: A process or set of rules followed by a computer to learn from data and make decisions or predictions. *4*

Menger sponge: A three-dimensional fractal curve. It is a cube that subdivides recursively into smaller cubes, with the center and the corners removed. This results in a highly porous object with an infinite surface area and zero volume in the limit. *171*

mutually assured destruction (MAD): A doctrine of military strategy where a full-scale use of nuclear weapons by two opposing sides would cause the complete annihilation of both. *39*

NAND Gate: A digital logic gate that produces an output which is false if and only if all its inputs are true. *109*

Nash equilibrium: An equilibrium situation in Game Theory in which a player will continue with their chosen strategy, having no incentive to deviate from it, after taking into consideration the opponent's strategy. *154*

nihilism: The philosophical viewpoint that suggests the lack of belief in meaningful aspects of life. *99*

NOR Gate: A digital logic gate that produces a true output if and only if all its inputs are false. *109*

paradox of competition: The idea that increased competition can sometimes lead to reduced overall competitiveness. *26*

parity bits: A simple form of error detecting code used in digital networks and storage devices to detect accidental changes to raw data. *45*

Parrondo's paradox: A paradox in game theory where two losing strategies combine to create a winning expectation. *84*

plebiscite: A direct vote by the entire electorate on an important public question, such as a change in the constitution. *51*

principle of explosion: A principle in classical logic where from a contradiction any statement can be proven. *107*

prosecutor's fallacy: A logical fallacy where the likelihood of a match (like DNA evidence) is equated with the likelihood of guilt, without considering other factors. *9, 79, 127*

Russell's paradox: A paradox which questions whether a set that contains all sets that do not contain themselves, contains itself. *174*

Schrödinger's cat: A thought experiment that illustrates the concept of superposition in quantum mechanics. *62*

shifting the burden of proof: A fallacy where the burden of proof is shifted to the person who is questioning or disproving a claim without having evidence for it. *100*

Simpson's paradox: A phenomenon in statistics where a trend appears in different groups of data but disappears or reverses when these groups are combined. *6, 7*

single transferable vote (STV): A voting system designed to achieve proportional representation through ranked voting in multi-seat organizations or constituencies. *54, 55*

slippery slope fallacy: A fallacious argument that suggests a minor action will lead to major and oftentimes ludicrous consequences. *102*

St. Petersburg paradox: A paradox in decision theory where, counterintuitively, only a few people would be willing to pay even a small amount to play a game, even though there is an expected infinite payoff. *32*

strawman argument: Misrepresenting someone's argument to make it easier to attack or refute. *107*

sunk cost bias/sunk cost fallacy: The tendency to continue an endeavor once an investment in money, effort, or time has been made. *83, 102, 189*

survivorship bias: Focusing on the surviving subjects of a process and overlooking those that did not survive, leading to false conclusions. *8, 86*

tactical voting: Voting for a candidate not because you support them, but to prevent another candidate from winning. *53, 55*

tautology: A statement or argument that is true by necessity or by virtue of its logical form. *135*

trolley problem: A thought experiment in ethics and psychology involving a moral dilemma of sacrificing one life to save others. *134*

Turing Labyrinth: A labyrinth pattern generated from a Turing type reaction-diffusion model. *85*

Turing test: A test of a machine's ability to exhibit intelligent behavior equivalent to, or indistinguishable from, that of a human. The test is passed if the evaluator, a human, cannot reliably tell the machine from the human. *156*

Will Rogers phenomenon: When moving an element from one set to another raises the average of both sets. *11*

XOR Gate: A digital logic gate that gives a true (or high) output if and only if the number of true inputs is odd. *109*

Zeno's paradox: A set of four paradoxes that deal with he counterintuitive nature of continuous space and time. *105*

zero-knowledge proof: A method in cryptography by which one party (the prover) can prove to another party (the verifier) that they know a value x, without conveying any information apart from the fact that they know the value x. *174*

zero-sum bias: The belief that a situation is like a zero-sum game, where one person's gain is another's loss. *77*

Zipf's law: An empirical law stating that in many types of data sets, the frequency of any word is inversely proportional to its rank in the frequency table. *44*

ABOUT THE AUTHORS

Inavamsi Enaganti, Director of the upcoming Param Science Centre, India, and alumnus of Chennai Mathematical Institute (CMI) and New York University (NYU), is fuelled by an insatiable curiosity. Having ventured boldly from his startup origins, crazy research, think tanks and NGOs, he aims to pursue the truth of reality. His first book at 15 years old, The Theory of Nonsense, started his journey to deciphering the world. He is driven by a profound desire to unravel the mysteries of reality and propel human civilization to its zenith within his lifetime. His heart beats to the rhythm of the cosmos — a longing not just to witness the craziness of human society or build a floating city but to pave a path for humanity to encounter alien civilisations. This isn't merely a job; it's his epic quest to truly enjoy life with the philosophy of YOLO.

Nivedita Ganesh is a budding young changemaker, a self-proclaimed woman of the future. By day she navigates the corporate jungle of business and banking. By night she's a music connoisseur, a classical dancer and cellist, moonlighting as an avid dystopian fiction reader. A maverick with strong academic roots who has lived in 4 different countries, she bridges the east and the west, as well as arts and science. She has a BSc (Hons) in Mathematics and Computer Science from Chennai Mathematical Institute (CMI) and Indian Institute of Technology Madras (IITM), and an MS in Computing, Entrepreneurship and Innovation from New York University (NYU).

Bud Mishra is a professor of computer science, mathematics, cell biology and human genetics at NYU, CSHL and MSSM. He is also an entrepreneur, inventor and mentor to young innovators and start-ups. Following the emergence of Covid pandemics, a large group of multinational and multidisciplinary team of technologists organized themselves as RxCovea under Bud's leadership; RxCovea remains active and trains young inventors how to solve Wicked problems using hypothesis driven minimal viable products and repeated pivoting as the MVPs fail cheaply and quickly and ultimately proceeding to scale. Bud has a degree in Science from Utkal University, in Electronics and Communication Engineering from IIT, Kharagpur, and MS and PhD degrees in Computer Science from Carnegie Mellon University. He is a fellow of IEEE, ACM and AAAS, a fellow of National Academy of Inventors (NAI), European Alliance for Innovation (EAI), a Distinguished Alumnus of IIT (Kharagpur), and a NYSTAR Distinguished Professor. He is currently focused on mitochondrial endosymbiosis and its role in aging, cancer and neurodegeneration, and nanomapping technology to assay mitochondrial heteroplasmy.

Note from the Illustrator

Alexander Lu: I'm sure you're wondering at this point why I chose to put so much time and energy into illustrating this math textbook. This, students, is a real life example of a **sunk-cost fallacy**, the phenomenon whereby a person is reluctant to abandon a strategy or course of action because they have invested heavily in it, even when it is clear that abandonment would be more beneficial. Ok, fine, I copied that definition off of the internet, but it still applies here. Plus, you probably used AI for your coding homework. What do you want from me? I'm not a writer.